Contents

Radiation and Health:
the Biological Effects of Low-level
Exposure to Ionizing Radiation

Radiation and Health: the Biological Effects of Low-level Exposure to Ionizing Radiation

Edited by

ROBIN RUSSELL JONES, MRCP
Chairman, Friends of the Earth Pollution Advisory Committee

and

RICHARD SOUTHWOOD, FRS
Chairman of the National Radiological Protection Board

A Wiley Medical Publication

JOHN WILEY & SONS
Chichester · New York · Brisbane · Toronto · Singapore

British Library Cataloguing in Publication Data:
Radiation and health: the biological effects
of low level exposure to ionizing radiation.
—(A Wiley medical publication).
1. Radiation—Physiological effect
I. Russell Jones, Robin II. Southwood,
Sir Richard
613 RA1231.R2

ISBN 0-471-91674-9

Typeset by Mathematical Composition Setters Ltd, Salisbury.
Printed and bound in Great Britain by Anchor Brendon Ltd, Tiptree, Colchester.

Preface

This volume represents the edited proceedings of an International Conference on 'The Biological Effects of Ionizing Radiation' held at the Hammersmith Hospital, London on 24 and 25 November 1986. The conference was chaired by Professor Sir Richard Southwood, FRS, Chairman of the National Radiological Protection Board, and organized by Dr Robin Russell Jones, MRCP, Chairman of the Pollution Advisory Committee to Friends of the Earth. The conference was sponsored by Friends of the Earth, UK and Greenpeace International. We would like to express our appreciation to both organizations.

In many ways the conference was unique in bringing together the main protagonists in the low-level radiation controversy. During the previous decade the debate about the health consequences of radiation exposure had become increasingly polarized and at times was in danger of becoming incoherent. This was particularly apparent in Britain where criticism of the nuclear industry had been focused by a series of public inquiries into proposals to build installations at Windscale (later renamed Sellafield), Sizewell and Dounreay. At the same time public concern had been aroused by reports of a higher than expected number of childhood leukaemias in the vicinity of certain nuclear establishments in the United Kingdom, particularly Sellafield and Dounreay.

Conventional radiobiological thinking does not allow for the possibility that discharges from nuclear facilities could account for such a large excess in the incidence of childhood cancers. Environmentalists argue that in the absence of any other identifiable agent, radiation is the most obvious explanation.

It was against this background that the conference was organized so that all sides involved in the debate could assess the validity of the data presented and the strength of the arguments put forward. Speakers were chosen on the basis of their current research involvement in the topic with particular reference to data which offered new insight into the subject. Sessional Chairmen were invited on the basis of their pre-eminence in the fields of radiobiology, epidemiology and environmental medicine, or their involvement in the low-level radiation debate. Those participating are listed on p. xiii–xv.

It was our intention throughout to organize a balanced programme which reflected equally the differing scientific viewpoints, and to ensure that the audience had an opportunity to participate in the debate. Because these discussions served both to clarify some of the research findings and to highlight the more important issues, they have been retained in this volume in edited form.

Whether or not the evidence allows a consensus to be drawn must be left for the readers themselves to decide. But certainly it is our hope that this volume will provide a rigorous, thoughtful and critical reappraisal of the evidence to date.

All but one of the papers given at the Conference has been included. The exception is Professor Wolfgang Jacobi who felt unable to furnish a written version of his oral presentation. Fortunately the data presented by Professor Jacobi has been reported elsewhere (Environmental Radioactivity and Radiation Exposure in Southern Bavaria from the Tschernobyl Accident. Report of the Institute for Radiation Protection, Munich-Neuherberg. *GSF Report 16/86* dated 15 June 1986).

Numerous people have provided generous help in the preparation of this volume. While we cannot thank them all we would particularly like to express our appreciation to the chapter authors who willingly and cheerfully met tight deadlines in order to bring this volume to completion as soon as possible. We owe a considerable debt of gratitude to the many secretaries who coped with a considerable volume of work and in particular Jacqueline Sawyer who transcribed the recorded proceedings. We would like to express our appreciation to the Dean and Secretary of the Royal PostGraduate Medical School who generously provided the venue and the facilities. Finally we are most indebted to Dr Gari Donn who fulfilled her task of Conference Secretary with both patience and efficiency.

ROBIN RUSSELL JONES and RICHARD SOUTHWOOD
March 1987

Contributors to this Volume

DR VALERIE BERAL, *Department of Epidemiology, London School of Hygiene and Tropical Medicine, London, WC1, UK.*

PROFESSOR ROGER BERRY, *Department of Oncology, Middlesex Hospital Medical School, London, W1, UK*

J. BLACK, *National Radiological Protection Board, Chilton, Didcot, Oxon, UK.*

MARGARET BOOTH, PhD, *London School of Hygiene and Tropical Medicine, London, WC1, UK*

LUCY CARPENTER, M.Sc., *London School of Hygiene and Tropical Medicine, London, WC1, UK.*

DR ROGER CLARKE, *Secretary, National Radiological Protection Board, Chilton, Didcot, Oxon, UK*

DAVID CROUCH, *Science Policy Research Unit, University of Sussex, Falmer, Brighton, Sussex, UK.*

DR SARAH DARBY, PhD, *Cancer Epidemiology and Clinical Trials Unit, University of Oxford, Radcliffe Infirmary, Oxford, UK.*

J. DIONIAN, *National Radiological Protection Board, Chilton, Didcot, Oxon, UK.*

RICHARD DOLL, DM, FRS, *Cancer Epidemiology and Clinical Trials Unit, University of Oxford, Radcliffe Infirmary, Oxford, UK.*

PROFESSOR H. J. EVANS, *MRC Clinical and Population Cytogenetics Unit, Western General Hospital, Edinburgh, UK.*

T. P. FELL, *National Radiological Protection Board, Chilton, Didcot, Oxon, UK.*

PATRICIA FRASER, MD, *London School of Hygiene and Tropical Medicine, London, WC1, UK.*

MISS FRANCES FRY, *National Radiological Protection Board, Chilton, Didcot, Oxon, UK.*

DAVID GEE, *General and Municipal Workers Union, Ruxley Ridge, Claygate, Esher, Surrey, UK.*

G. W. KNEALE, Regional Cancer Registry, Queen Elizabeth Medical Centre, Birmingham, UK.

DR BARRIE LAMBERT, *Department of Radiation Biology, Medical College of St Bartholomew's Hospital, Charterhouse Square, London, EC1, UK.*

DR PAUL LEWIS, FRCPath., *Department of Histopathology, Royal Postgraduate Medical School, Hammersmith Hospital, Du Cane Road, London, W12, UK.*

PROFESSOR KARL Z. MORGAN, *Former Director, Health Physics Division, Oakridge National Laboratory, USA.*

C. R. MUIRHEAD, *National Radiological Protection Board, Chilton, Didcot, Oxon, UK.*

PROFESSOR EDWARD RADFORD, *Professor of Epidemiology, University of Pittsburgh, USA.*

PROFESSOR SIR RICHARD SOUTHWOOD, FRS, *Chairman, National Radiological Protection Board (NRPB), Chilton, Didcot, Oxon, UK.*

DR JOHN STATHER, *National Radiological Protection Board, Chilton, Didcot, Oxon, UK.*

DR ALICE STEWART, *Regional Cancer Registry, Queen Elizabeth Medical Centre, Birmingham, UK.*

DR PETER TAYLOR, *Director, Political Ecology Research Group (PERG), 34 Cowley Road, Oxford, UK.*

PROFESSOR Y. UJENO, *Department of Experimental Radiology, Faculty of Medicine, Kyoto University, Kyoto 606, Japan.*

JOHN URQUHART, *Editor, Journal of Nuclear Risks, Low Gosford Home Farm, Bridle Path, Gosforth, Newcastle Upon Tyne, UK.*

Contributors to the Conference

The following contributed to the 'International Conference on the Biological Effects of Ionizing Radiation' held at the Hammersmith Hospital on 24 and 25 November.

LIST OF CHAIR PERSONS

Conference Chairman

PROFESSOR SIR RICHARD SOUTHWOOD, FRS,
Chairman, National Radiological Protection Board, Former Chairman of the Royal Commission on Environmental Pollution.

Sessional Chair Persons

SIR DOUGLAS BLACK FRCP,
Chairman of the Independent Advisory Group on Cancer in West Cumbria.

PROFESSOR MARTIN BOBROW FRCP,
Prince Philip Professor of Paediatric Research, Guy's Hospital, Chairman of the DHSS Committee on the Medical Aspects of Radiation Exposure (COMARE).

SIR RICHARD DOLL, DM, FRCP, FRS,
Epidemiological Consultant to the National Radiological Protection Board.

PROFESSOR J. H. EDWARDS, FRCP, FRS,
Professor of Genetics, Oxford University.

D ROBIN RUSSELL JONES, MRCP,
Chairman, Friends of the Earth Pollution Advisory Committee.

LIST OF SPEAKERS

DR VALERIE BERAL,
Department of Epidemiology, London School of Hygiene and Tropical Medicine, London, WC1, UK.

PROFESSOR ROGER BERRY,
Department of Oncology, Middlesex Hospital Medical School, London, W1, UK.

DR ROGER CLARKE,
Secretary, National Radiological Board, Chilton, Didcot, Oxon, UK.

DAVID CROUCH,
Science Policy Research Unit, University of Sussex, Falmer, Brighton, Sussex, UK.

DR SARAH DARBY, PhD,
Cancer Epidemiology and Clinical Trials Unit, University of Oxford, Radcliffe Infirmary, Oxford, UK.

PROFESSOR H. J. EVANS, MRC,
Clinical and Population Cytogenetics Unit, Western General Hospital, Edinburgh, UK.

MISS FRANCES FRY,
National Radiological Protection Board, Chilton, Didcot, Oxon, UK.

DAVID GEE,
General and Municipal Workers Union, Ruxley Ridge, Claygate, Esher, Surrey, UK.

PROFESSOR WOLFGANG JACOBI,
Institut fur Straplerschultz Irgolstadter, Landstrasse 1, D-8042 Neuherburg, West Germany.

DR BARRIE LAMBERT,
Department of Radiation Biology, Medical College of St Bartholomew's Hospital, Charterhouse Square, London, EC1, UK.

DR PAUL LEWIS, FRCPath,
Department of Histopathology, Royal Postgraduate Medical School, Hammersmith Hospital, Du Cane Road, London, W12, UK.

PROFESSOR KARL Z. MORGAN,
Former Director, Health Physics Division, Oakridge National Laboratory, USA.

PROFESSOR EDWARD RADFORD,
Adjunct Professor of Epidemiology, University of Pittsburgh, USA.

PROFESSOR SIR RICHARD SOUTHWOOD, FRS,
Chairman, National Radiological Protection Board (NRPB), Chilton, Didcot, Oxon, UK.

DR JOHN STATHER,
National Radiological Protection Board, Chilton, Didcot, Oxon, UK.

DR ALICE STEWART,
Regional Cancer Registry, Queen Elizabeth Medical Centre, Birmingham, UK.

DR PETER TAYLOR,
Director, Political Ecology Research Group (PERG), 34 Cowley Road, Oxford, UK.

PROFESSOR Y. UJENO,
Department of Experimental Radiology, Faculty of Medicine, Kyoto University, Kyoto 606, Japan.

JOHN URQUHART,
Editor, Journal of Nuclear Risks, Low Gosford Home Farm, Bridle Path, Gosforth, Newcastle Upon Tyne, UK.

Part 1
Introduction

Radiation and Health
Edited by R. Russell Jones and R. Southwood

1

Opening Remarks

SIR RICHARD SOUTHWOOD FRS
(*Conference Chairman*)

The purpose of this conference is to explore questions on the effects of ionizing radiation on biological systems—more particularly—the effects of low-level radiations on man. In organizing this meeting it has not been the intention of Dr Russell Jones and myself to put forward a programme which will project a uniformity of view; rather we have sought to explore areas where disagreements have frequently been voiced.

Ionizing radiations are of course part of our natural environment and life has been exposed to them from the very beginning of evolution. Relatively recently, virtually within living memory, they have been harnessed to the service of man, firstly in medicine and secondly in industry. In both areas great benefits were initially perceived—but slowly it became apparent that, as with so many opportunities, they offered something of a two-edged sword. If you find you have a two-edged sword it is not always necessary to abandon it, but one must perceive its dangers and determine if it can be used in a way so that harm may be avoided. There are two processes in such an assessment, firstly an evaluation of the risk of harm in relation to different modes of operation, and secondly the determination of what is an acceptable risk.

This conference is about the recognition of the risk: precisely about the dose/risk relationship for ionizing radiation and various malignant diseases of man. The conference is *not primarily* about the determination of what is an *acceptable risk*, nor will we be discussing how one balances the risk that arises from, for example, nuclear power, against the risks associated with other forms of energy generation and other forms of human endeavour and how such a balance takes account of man's imperative need for energy—nor of how a clinician balances the diagnostic value of an X-ray against its mutagenic risk.

As I said at the outset we are all, of course, exposed to ionizing radiation and Figure 1 gives an idea of the relative contributions from different sources. The accident at Chernobyl occurred during the last year and one can see from this diagram how the contribution that it made to exposure in the UK was

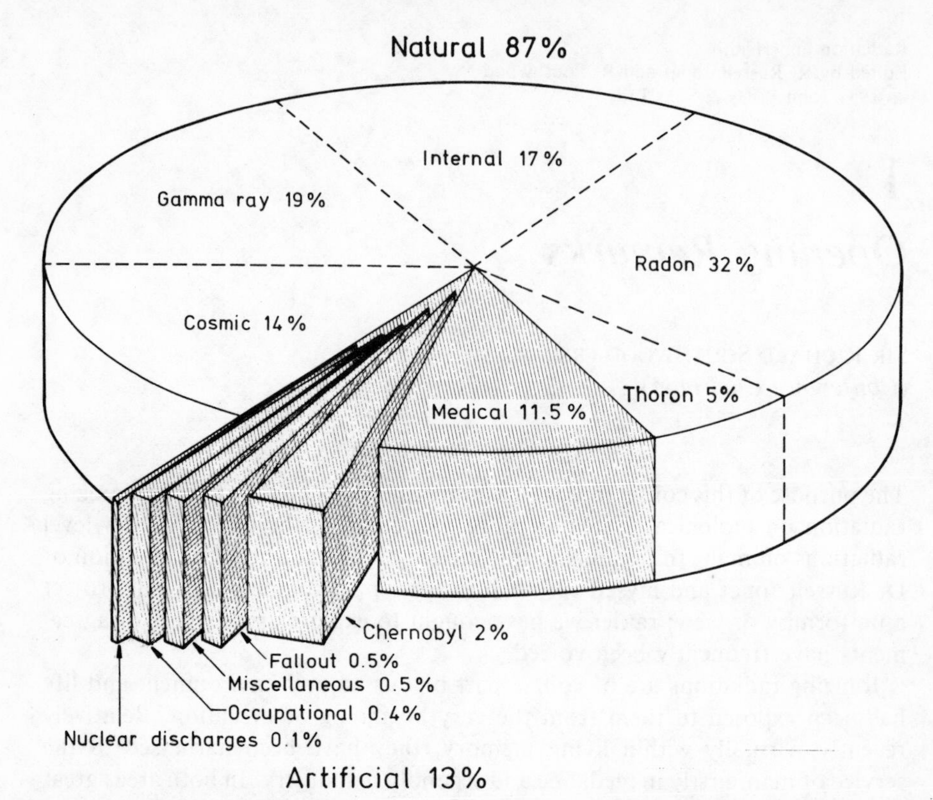

Figure 1 Radiation exposure of the UK population; contributions to the average effective dose equivalent. (*Note*: The contribution from the Chernobyl accident is shown as an *additional* 2 per cent, so that the artificial component is now 15/102*ths*).

small relative to the total amount of ionizing radiation, but was apparently very large in relation to the amount that we received from our own nuclear industry.

It is of course very much easier to measure the amount of radioactivity in the environment than to determine its effect on human bodies, for the use of the experimental method is clearly very limited in such investigations. Basically, the biological effects of ionizing radiation have been determined by relating effects to dose. Unfortunately there are problems for both parameters. The ill-effects are of the type where there are several causative agents. Until recently, the information on total personal dosimetry was relatively modest and, as we will hear, there is still some uncertainty about the doses of some forms of radiation on the target cells. A putative relationship is most easily determined when the dose has been relatively high, either in respect of some medical treatment or to survivors from nuclear weapons. As can be seen from Figure 2, such information provides a cloud of points for high exposure and

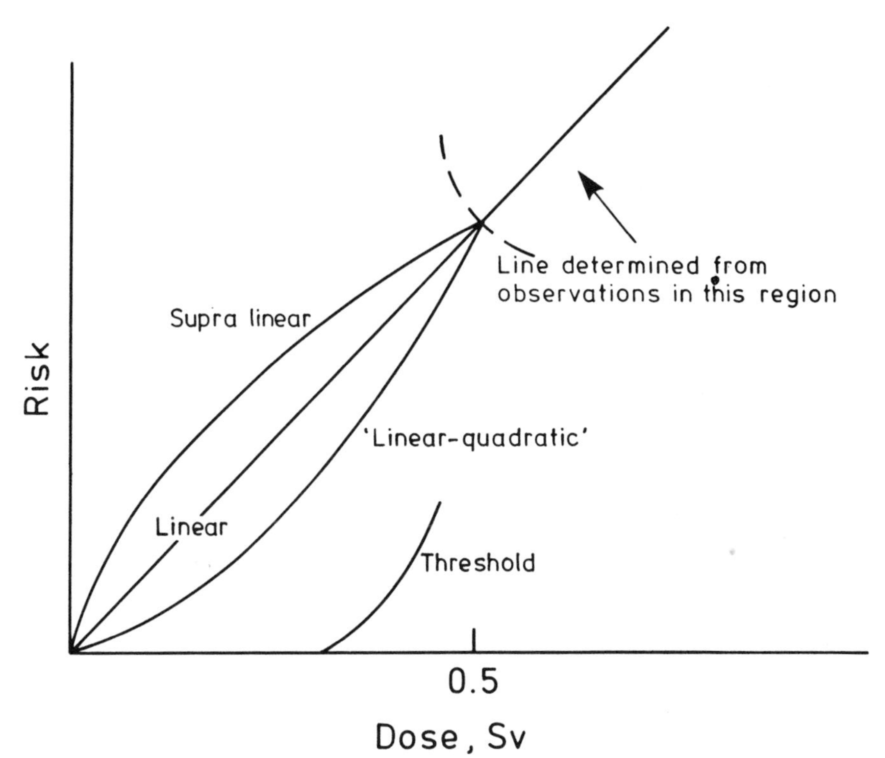

Figure 2 Various forms of the dose–risk relationship.

a reasonably reliable line can be calculated for the risk. The problem arises in determining what happens at lower levels. Should the line be projected as a straight linear line to zero or is there some non-linear relationship of enhanced or reduced sensitivity at lower levels or perhaps, related to the body's repair mechanisms, a threshold? The form of this line for low levels is one of the important questions that this conference has to consider. Another is the extent—if any—to which the form of the line varies under different conditions, both physical and biological, i.e. with different types of radiation, persons of different ages.

I will not explore these matters further—many speakers will do so.

There is, however, one other general point that I would like to make. It is to draw your attention to the two approaches used for the analysis of a complex ecological system—when it is not possible to carry out detailed experimental replication. The two approaches are shown in Figure 3. The first is to investigate the system in a step-wise fashion. The various interactions are identified and constitute a chain of processes. Predictions may be made for a quantity at any particular step and such predictions can be tested against

PUTATIVE CAUSE AND EFFECT CHAINS

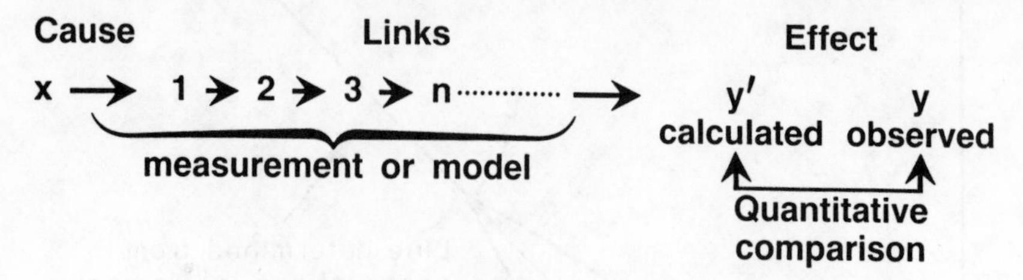

EXTENSIVE CORRELATIVE STUDY

Cause		Correlations		Effect
x_1	....	r_1	.... →	y_1
x_2	....	r_2	.... →	y_2
x_3	....	r_3	.... →	y_3
x_n	....	r_n	.... →	y_n

Figure 3 Two approaches to the analysis of complex cause and effect systems.

measurements made in the field. In our system such models are used to predict a dose experienced under certain conditions and using the dose/risk relationships that I have referred to above, predictions are made of the risk which can then be compared with data obtained from actual populations. This given a check on the model and the dose/risk relationship. The alternative approach, that in ecology is often referred to as the extensive approach, is essentially an epidemiological one. Patterns of the putative cause and effect can be compared. Everyone is well aware of the dangers of false correlations and therefore there is need not only to replicate, but to apply careful statistical tests to such analyses. Several such epidemiological approaches will be presented to us in the next two days. The greatest confidence can be accorded to any quantitative theory that is supported by both these approaches. If they appear to give different indications then further research and further analysis are necessary.

Part 2

Environmental Issues

Radiation and Health
Edited by R. Russell Jones and R. Southwood

Published 1987 by John Wiley & Sons Ltd.

2

Doses from Environmental Radioactivity

F.A. FRY
National Radiological Protection Board,
Chilton, Didcot, Oxon., UK.

ABSTRACT

Radionuclides are widespread in our environment. Most occur naturally and are either of primordial origin or are continually produced by natural processes. Radionuclides produced by anthropogenic sources make a small additional contribution to environmental levels but are subject to intensive monitoring programmes and, perhaps inevitably, attract a disproportionate share of public interest.

The aim of this chapter is to discuss the sources of radioactivity in the environment and the routes by which the population is exposed. The scope of measurement programmes within the United Kingdom is briefly described. Finally, doses from the various sources are evaluated and compared.

INTRODUCTION

For some years there has been considerable public concern about radiation in general and man-made radioactivity in particular. Public awareness of the impact of radionuclides in the environment has been heightened even more by the Chernobyl reactor accident. This conference is an opportune occasion at which we can review sources of exposure and discuss potential biological effects of radiation. Many of the following chapters will present doses and risks to particular groups of the public or workers. The aim of this chapter is to provide a review of radioactivity in the environment and to give an appreciation of the doses to which we are all exposed. Thus it may provide a framework within which other exposures can be placed in perspective.

SOURCES OF EXPOSURE

Radiation of natural origin is widespread in the whole environment. Radiation reaches earth from outer space, the earth itself is radioactive and naturally-occurring radionuclides are present in the air we breathe, in the food we eat and in our own bodies. Everyone is exposed to natural radiation and for most people it is the highest contributor to total dose.

Man-made radionuclides have been distributed throughout the world as a result of nuclear weapons testing in the atmosphere. These radionuclides are inhaled, they are deposited on the ground giving rise to external exposure and they can be transferred through food-chains to our diet. Even though the period of intensive weapon testing occurred more than twenty years ago, residual activity from these tests and from occasional more recent explosions still give rise to some small exposure of the population.

Radioactive materials are discharged from nuclear installations, from other industrial premises and from medical and research institutes. All discharges must be authorized and monitoring programmes are prescribed for significant discharges. Operators of nuclear sites monitor the activity before it is discharged and they undertake environmental monitoring programmes. Check measurements, both on discharges and in the environment, are made by the authorizing departments. Organizations, such as medical and research institutions, that discharge small quantities of radionuclides do not normally carry out routine monitoring programmes although in some circumstances the radionuclides they discharge can be readily measured in the environment.

Accidental releases of radionuclides may also occur and, as Chernobyl has shown, a severe accident at a nuclear power station can lead to widespread contamination of the environment.

ROUTES OF EXPOSURE

Radionuclides are subject to all the physical, chemical and biological processes of environmental transfer. No matter how complex the pathway by which the activity may reach man, the actual routes of human exposure are limited to: external irradiation; inhalation of airborne material; ingestion of activity in food or water. Measurements are related to those mechanisms of exposure. Thus, in the environment we measure external dose rates and activity concentrations in air, water and foods. In exceptional circumstances, radionuclides can be measured directly in the human body.

The quantity we cannot measure is the one we require—dose. Thus, whatever measurements are made, some calculation must be performed to arrive at doses to persons. To put doses from all sources on a common basis, the quantity we require is the sum of effective dose equivalent from external irradiation in a year and the committed effective dose equivalent from intakes of radionuclides during that year. The term committed refers to the dose from intake of a radionuclide integrated over a person's lifetime, taken conventionally to be 70 years. For simplicity, we shall use the term dose, unless clarification is necessary. Even in the simplest case of external irradiation, it is necessary to convert the instrument response to dose in the body and we need to know how long persons spend in the location to calculate the actual dose received from that particular source. In the somewhat more complex situation of intakes of radionuclides, we need to know breathing rates and food consumption rates; we need to know how the radionuclides behave in the body; and we need to calculate the energy absorbed in body tissues over time and from that the dose.

In the case of most nuclear installations, routine discharges do not lead to measurable dose rates or activities in the environment. Doses must therefore be calculated from a knowledge of the quantities and forms of the radionuclides discharged, their dispersion in the environment and their subsequent fate.

We now consider two sources of public exposure: natural radiation and artificial radionuclides that are widespread in the environment. We shall present average doses for the United Kingdom but, where applicable, we shall also give an indication of the range of doses.

NATURAL RADIATION

The main components of natural radiation are: cosmic radiation; terrestrial gamma-rays; radon decay products; other radionuclides in air, food and water. In recent years, NRPB staff have undertaken an extensive programme of measurements in relation to natural radiation. Here it is possible to draw on only a few results of this work in connection with public exposure.

Outdoor exposure from external irradiation is due almost entirely to cosmic rays and terrestrial gamma-rays. In the United Kingdom, the annual dose from cosmic rays is about 300 μSv, on average.[1] Cosmic radiation varies with altitude and latitude. The annual dose varies from about 280 μSv a year in the south of England to 310 μSv a year in the north of Scotland. Since most of the population lives at low altitude, there is little variation in cosmic ray dose with altitude.

Terrestrial gamma-rays are emitted by radionuclides in the uranium and thorium series and by potassium-40, all of which are present in the earth's crust at concentrations which depend upon the type of rock and soil. We have measured gamma-radiation out of doors throughout Great Britain.[2] The mean value for the country is about 240 μSv a year. However, most people spend most of their time indoors—75 per cent at home, with a further 15 per cent in other buildings.[3] We have undertaken a study of the national indoor exposure by measurements in some 2000 homes chosen randomly.[4] The measurements were made with two small detectors, a thermoluminescence device for gamma-radiation and a passive detector for the radon-222 concentration. The mean annual dose from gamma-radiation during occupancy of a dwelling is about 380 μSv a year, giving a total annual dose from

Table 1 Annual dose in the UK from radiation of natural origin (μSv)

Source	Average	Range
Cosmic radiation	300	280–310
Terrestrial gamma-rays	400	200–1000
Radon and thoron decay products	800	100–10 0000
Other internal radiation	370	200–400
Total	1870	1000–100 000

gamma-radiation of about 400 μSv. The range of values is approximately 200–1000 μSv.

When radon or thoron gas enters the atmosphere from the ground, it is dispersed in the air and concentrations outdoors are low. However, when these gases enter a building, either from the walls or through the floor, the concentrations build up because of the restriction in the supply of outdoor air. The immediate decay products are radionuclides with short half-lives which become attached to dust particles in the air and can be inhaled. Our national survey includes measurements of radon-222 and we have also conducted more detailed surveys in areas where the local geology indicated that elevated exposures might occur.[5] When we include exposure outdoors and in buildings other than dwellings, the average dose from radon and thoron decay products is 800 μSv in a year, but there are considerable variations about this value.[1] It is worth noting that there are some dwellings in which the occupants are receiving doses of 100 mSv a year, that is, more than the annual dose limit for occupationally exposed workers.

Other radionuclides from the uranium and thorium series are present in air, food and water. Their contribution to average dose is conventionally taken as 170 μSv a year although a value of 90 μSv may be more appropriate for UK diets.[6] The other major contributor to dose is potassium-40 which can be measured directly in the human body. The level of potassium in the body is controlled biologically; it varies with age, sex and state of health. The range of doses for adults of both sexes is probably 110 μSv to 210 μSv with a typical value for a healthy adult male of 180 μSv a year. Other radionuclides in diet make smaller contributions, bringing the total dose from intake to 370 μSv in a year with a probable range of 200 μSv to 400 μSv.

Doses from all natural sources are summarized in Table 1.

ARTIFICIAL RADIONUCLIDES IN THE ENVIRONMENT

The Board operates an environmental radioactivity surveillance scheme to measure concentrations of radionuclides in various media and to evaluate dose to the population from widespread artificial radionuclides. A network of air and rain sampling stations is operated in parallel with the Atomic Energy Research Establishment at Harwell (AERE), but the Board also measures radionuclides in milk as a guide to activity in diet. Until May of this year, the radionuclides detected were generally naturally-occurring or produced as a result of nuclear weapons testing in the atmosphere.[7] The reactor accident at the Chernobyl nuclear power station in the western Ukraine on 26 April released substantial quantities of radionuclides to the atmosphere, eventually resulting in widespread contamination throughout the whole of the northern hemisphere and leading to substantial increases in measured values.[8]

During 1984, the levels of weapons fallout radionuclides in airborne dust and in deposition were the lowest reported since measurements were commenced in 1953 by AERE.[9] Similarly, the concentrations of caesium-137 and strontium-90 in milk were the lowest measured.[7] The average dose is shown in Table 2. The dose for external irradiation is estimated from the measured cumulative deposition of activity and models of indoor and outdoor exposure. The dose from carbon-14 is calculated from the excess activity per unit mass of this radionuclide in the biosphere, from reference dietary intake and standard metabolic and dosimetric models. For strontium-90 and caesium-137, the doses are based on measured concentrations in milk with an extrapolation to total diet and again on standard metabolic and dosimetric models. The average dose in 1984 was about 7 μSv, compared with a value of about 80 μSv in the early 1960s.[1] Currently the range of doses within the United Kingdom is perhaps 6 μSv to 10 μSv. Levels and doses for 1985 are similar to those for 1984, and, in the absence of a further input, were expected to continue to decline, but only slowly.

The accident at the Chernobyl reactor will be dealt with in a later session, but no discussion of doses from environmental radioactivity would be complete without at least a brief mention of this. Persons were exposed by several mechanisms: immersion in the cloud; inhalation of activity; irradiation by deposited material; ingestion of radionuclides. In the intermediate and long-term, the radionuclides of radiological significance are the caesium isotopes since these are transferred through food chains to our diet. As a result of the

Table 2 Average annual dose from weapons fallout (μSv)

External irradiation	1.4
Intake	
Carbon-14	3.8
Strontium-90	1.4
Caesium-137	0.3
Total	6.9

accident, the concentration of caesium-137 in airborne dust reached a value of almost 1 Bq m^{-3}, considerably higher than any concentration measured during the period of intensive weapons testing, but, since the activity was injected at low elevation in the troposphere, this level persisted for only a short time in contrast to the stratospheric fallout from weapons tests.

The deposition of activity was dependent upon meteorological conditions at the time of the passage of the contaminated air mass. High deposition occurred in those parts of the United Kingdom where it rained heavily during the passage of the cloud. These include north Wales, the Lake District and south-west Scotland. Deposition is measured either by collecting rainfall or, at later stages, by taking soil cores. In the 'wet' regions, the deposition of caesium-137 from this accidental release now predominates over the weapons fallout level but, in areas where it did not rain, dry deposition of activity has made a negligible addition to cumulative levels from weapons testing. Concentrations of caesium-137 in milk reached peak values of 40 Bq l^{-1} in the United Kingdom as a whole, 200 Bq l^{-1} in the wet areas, compared with typical values of 0.06 Bq l^{-1} before the accident. Caesium isotopes can be measured in other foodstuffs and the results used to infer dose from diet, but the activity can also be measured in the human body and, in the long term, the *in vivo* measurements will provide a basis for refining dose estimates from contamination of diet with these radionuclides.

Our current estimates of dose in the United Kingdom from all mechanisms of exposure as a result of the Chernobyl reactor accident are given in Table 3. The average value for all years is 50 μSv, but the dose to persons resident in areas of high rainfall may be a factor of five higher. These values are less than the estimates given in a preliminary evaluation of doses in the UK.[10] The difference in the values arises for two main reasons. Firstly, deliberately cautious assumptions were made in the early paper. Secondly, more detailed information that was not available early in May has permitted better delineation of areas of high deposition and these are now known to affect fewer people than originally estimated.

Table 3 Typical dose to adults in the UK as a result of the Chernobyl reactor accident (μSv)

	First year	All years
England	25	30
Wales	70	95
Scotland	90	160
Northern Ireland	100	170
UK	35	50
Wet areas	190	270

Table 4 Annual dose from discharges (μSv)

	Average	Range
To sea, fish and shell-fish	1.3	0–1000
Other pathways	0.1	—
To atmosphere	0.1	—
Total	1.5	0–1000

Finally, we consider the dose from discharges from the nuclear industry. The most important pathway is liquid discharges leading to contamination of fish and shell-fish. The dose to the population as a whole is estimated by the Directorate of Fisheries Research from data on fish and shell-fish landings from relevant sea areas and measurement of average concentrations of fission products and actinides in fish and shell-fish in these areas.[11] The collective dose to the UK population in 1984 was 70 man Sv. These doses are not distributed uniformly over the population; they range from zero, for those individuals who do not consume fish and shell-fish, to around 1000 μSv for avid consumers in Cumbria. If we were to average this dose over the whole population, the result would be 1.3 μSv (Table 4). Discharges of other radionuclides do not lead to measurable concentrations and thus doses are calculated from models of dispersion, depletion and accumulation in water bodies followed by an analysis of ingestion, inhalation and external irradiation.[12] On average, these would add almost 0.1 μSv.

Discharges to atmosphere do not give rise to measurable levels of activity in the environment and, again, we must resort to calculations of dispersion, deposition and subsequent transfer through the terrestrial environment, followed by analysis of inhalation, ingestion and external irradiation.[12] This source of exposure adds a further 0.1 μSv to the average dose.

SUMMARY OF DOSE FROM ALL SOURCES

We can now compare doses from all sources of exposure (Table 5). The use of radiation and radionuclides in medical diagnosis and therapy, although of direct benefit to the patient, makes a significant contribution to the average dose to the population. It is the largest man-made source of exposure. The distribution of medical doses is obviously not uniform; many persons will, for many years, receive no dose at all from this source. The average dose to occupationally-exposed persons is 1400 μSv in a year; averaged over the whole population, this makes a small contribution of 8 μSv a year. Finally, most members of the public receive some exposure from a number of sources such as luminized watches, television receivers and air travel; such miscellaneous sources contribute 11 μSv a year on average.

Table 5 Average dose in the UK (μSv)

Source	Average
Environmental sources	
Natural radiation	1870
Weapons fallout	7
Discharges	1.5
Chernobyl accident[a]	50
Other sources	
Medical procedures	250
Miscellaneous sources	11
Occupational exposure	8

[a] The value for the Chernobyl accident is for all time. Other values are annual doses.

There is no doubt that natural radiation is the dominant source of exposure and the variations in this contribution are much larger than the total dose from man-made sources. For the current year, the Chernobyl reactor accident is the most important source of dose from artificial environmental radioactivity, but even this is not as high as the dose received from weapons testing in the 1960s. Weapons fallout, the Chernobyl reactor accident and controlled discharges together account for only a few percent of the total dose we all receive. This does not imply that such sources are justified, nor that we can neglect them, but it does perhaps give a sense of proportion to public exposure as a whole.

ACKNOWLEDGEMENTS

I am most grateful to my colleagues for the data on doses from natural radiation.

REFERENCES

1. J. S. Hughes and G. C. Roberts, The radiation exposure of the UK population—1984 review. *NRPB-R173* (1984).
2. E. J. Bradley and B. M. R. Green, Outdoor gamma-ray dose rates in Great Britain—preliminary results. *Radiol. Prot. Bull.*, **59**, 16–19 (1984).
3. L. Brown, National radiation survey in the UK: indoor occupancy factors. *Radiat. Prot. Dosim.*, **5**, 203–8 (1984).
4. B. M. R. Green, L. Brown, K. D. Cliff, C. M. H. Driscoll, J. C. M. Miles and A. D. Wrixon, Surveys of natural radiation exposure in UK dwellings with passive and active measurement techniques. *Sci. Tot. Env.*, **45**, 459–66 (1985).
5. K. D. Cliff, A. D. Wrixon, B. M. R. Green and J. C. H. Miles, Radon and its decay-products: occurrence, properties and health effects, ed. P. K. Hopke. Washington, American Chemical Society, in press.
6. J. L. Smith-Briggs, and E. J. Bradley, Measurement of natural radionuclides in UK diet. *Sci. Tot. Env.*, **35**, 431–40 (1984).

7. D. M. Smith, G. McAllister, D. Welham and D. Orr, Environmental Radioactivity Surveillance Programme: results for the UK for 1984. *NRPB-R189* (1985).
8. M. C. O'Riordan, and F. A. Fry, Cloud over Britain: dose from Chernobyl. Presented at the meeting of the British Association for the Advancement of Science, Bristol, 1986.
9. R. S. Cambray, K. Playford and G. N. J. Lewis, Radioactive fallout in air and rain: results to the end of 1984. *AERE R11915* (1985).
10. F. A. Fry, R. H. Clarke, and M. C. O'Riordan, Chernobyl reactor accident: early estimate of radiation doses in the UK. *Nature*, **321**, 193–5 (May 1986).
11. G. J. Hunt. Radioactivity in surface and coastal waters of the British Isles 1984. *Aquatic Environment Monitoring Report Number 13*. Ministry of Agriculture, Fisheries and Food, Directorate of Fisheries Research, 1985.
12. G. Lawson, N. McColl, J. A. Jones, J. Williams, J. Dionian, and J. R. Cooper, The radiological impact of routine discharges during 1970 to 1989 for the UK civil nuclear power programme. To be published.

Radiation and Health
Edited by R. Russell Jones and R. Southwood

3

The Interpretation of Monitoring Results

P. J. TAYLOR
Political Ecology Research Group, Oxford, UK.

INTRODUCTION

It is clear from the level of public controversy surrounding reports of radioactivity in the environment arising either from accidents at nuclear installations, or from routine planned discharges, that monitoring results can be subject to widely differing interpretations. These differences are not simply divided between scientists and a lay public subject to irrational fears, as some would believe, but also permeate the scientific establishment itself, where in certain instances fierce controversy has arisen over appropriate discharge levels or actions to be taken following accidental releases.

Such scientific controversies are seldom aired in the journals concerned with that science, perhaps being thought more appropriate for sociological disciplines. The professions concerned are often left with a sense of solidarity around their interpretations, with 'dissidents' relegated to various categories of incompetence or 'political' motivation. The treatment of Bowen's evidence to the Windscale Inquiry[1] provides an example of how the UK authorities contested and marginalized Bowen's concern for human health and his predictions of the eventual fate of actinides in sediment (reviewed by Stott Taylor[2]).

There are many more recent examples of monitoring results reported which have caused public concern at levels which would be dismissed by many established authorities in the radiological professions: Chernobyl is a case in point and illustrates the extent to which public confidence in 'government' science could reach a low ebb. It has been strongly eroded by the events surrounding the reprocessing plant at Sellafield, the major source of radioactive pollution to the European environment. On the one hand the authorities concerned with the protection of the public have been consistently assuring the public that the situation was, variously, 'of no concern', 'insignificant', 'negligible' and 'well within internationally accepted guidelines or limits'. Yet within the past few years 25 miles of beaches have been closed following a spill, a high-level scientific inquiry (the Black Committee) has investigated and confirmed an abnormally high incidence of childhood leukaemia in the Sellafield area (following data gathered by an independent television research team), and the company concerned has been forced to spend upwards of £200 million on discharge controls. How do these events and concerns relate to the oft-quoted pie diagram of nuclear power activities only increasing the environmental doses of the average Briton by less than 1 per cent? Is there just cause for concern at a scientific level, or are the professionals, who really understand, being beleaguered by a fearful, irrational and ignorant public, not a little misled by politically motivated dissident scientists?

As with all controversy, the answers are likely to lie somewhere in the middle, and this chapter is an attempt to chart that middle course, looking at those areas where there may be cause for concern (i.e. damage to the environment and/or human health) backed by scientific judgement, and also attempting to explain why, even when the science may provide reasonable grounds for assurance, this may be far from saying there is no 'environmental' impact.

THE SCIENCE OF RISK ASSESSMENT

That there is a science of risk assessment is peculiarly little appreciated by the protagonists in debates about low levels of radioactive contamination. It is a discipline that has been regularly applied to the field of major hazard assessment, such as that presented by oil, gas and nuclear power industries where rare events may have catastrophic consequences, and also to comparative assessments, where a whole fuel cycle is compared with another, for example, nuclear with coal, for its health or environmental impacts. The tools that have been developed in these assessments are directly relevant to the low-level radiation debate (for a review of risk assessment, see Ref. 3).

The first and foremost principle in risk assessment is the separation of two processes: risk estimation and risk evaluation.

Risk estimation

This is a process whereby the risk is expressed in numerical terms. This is the stage at which 'objective' science plays the main role. However, the science can never be entirely 'pure', for there are often uncertainties in the science where 'judgement' may be used to bridge inadequacies in data or models. Furthermore, there is a non-objective selection process that operates in the choice of appropriate data, design of experiments or research programmes from which data may be gained, as well as the constraints set upon the analysis by the necessity for numerical indices. That which can be estimated numerically is not always that which is of most concern and there are legitimate impacts which are not amenable to quantification (for example, compare the quantifiable health risks from unsterilized rubbish, with the unquantifiable aesthetic impact which would still accrue following complete sterilization). This example was put to the Holliday Committee on the Sea Dumping of Nuclear Waste, where dumping of 'scientifically' innocuous barrels (disputable) could be as much an affront to those who valued the 'sanctity' of the ocean as dumping sterilized rubbish in the grounds of Salisbury cathedral with the plea that it would harm nobody. Holliday accepted the reality of social impact and recommended the development of analytical procedures for environmental assessments.[4]

Risk evaluation

Let us suppose that agreement is reached on the 'objective' size of the risk. Let us assume for example, that the risk to a Danish fish consumer was 1 in 100 million per annum of contracting cancer from the Sellafield discharges. Is that risk acceptable? How does it compare with other risks? Does it matter how it compares? If we are to compare it, is it appropriate to compare to risks of falling off ladders in the home, smoking cigarettes, or being struck by lightning? If after we, or, say, a Committee of the International Commission on Radiological Protection, has compared it and found it 'negligible', what happens if the Danes disagree, complain to their Parliament, and their Environment Minister sends a diplomatic note to the UK Prime Minister?

The latter case is an historic event. What transpired in UK government is, of course, not documented, but we can suppose that the Prime Minister was advised that the Danish concern was purely political, arose from irrational public fear, had no 'scientific' basis, but could not be entirely ignored, for political reasons. The UK government then issues a statement that despite having met all its international obligations, it was continuing to improve the discharge situation. On another occasion, the annual sea dump in the North-East Atlantic, the government chose to defy the verdict of the London Dumping Convention, and was only prevented from continuing with marine disposal of nuclear waste by a trade union boycott. In this case, monitoring had established that the drums were leaking but no radioactivity had reached the surface from the 4 km depth.[5]

The methodology of risk evaluation has developed several cardinal principles relating to evaluation:

(i) That like must be compared with like: for example a physical accident causing death is not comparable with a death caused by cancer or hereditary defect. The two are perceived very differently and, involve different 'manner' of death.

(ii) A death or injury arising from accidents in which the person had a degree of personal responsibility or control is different to one whereby the 'locus of control' is outside that person's influence (as is the case with widespread pollution, and clearly not the case with cigarettes or motoring).

(iii) The death of a child from leukaemia is not equivalent in impact to the death of a person at age 70 from this disease.

There are several more elaborations of the nature and timing of death and injury which make simple numerical indices a minefield for the would be analyst seeking to compare or evaluate a risk. At the end of the day, public risks are strongly affected by the perceived benefits from an activity (nuclear electricity, reprocessing, etc.) and the perceived alternatives to that activity. It should be clear that the word ‘acceptable’ carries an enormous loading of value judgements, yet this is precisely the word used by ICRP. These matters are reviewed at length in the Political Ecology Research Group submission to the Sizewell Inquiry in a document comparing the health risks from the coal fuel cycle to those from the nuclear fuel cycle.[6]

THE ACCEPTABLE DOSE?

The ICRP has promulgated a system of dose limitation to be applied to public exposure from the nuclear fuel cycle. The system is described elsewhere in this symposium, and to briefly summarize, it hinges upon the concept of a dose level which is considered (by them) to be ‘unacceptable’, and below which doses must be kept As Low As Reasonably Achievable (ALARA). Wordings have changed since the first limits were proposed, and the upper limit is either 5 mSv (500 mrem)/per annum for short-term exposure (a few years) or 1 mSv/per annum for longer-term pollution likely to remain for a lifetime. The latter exposure would produce an additional risk of between 1 : 100 000 to 1 : 10 000 of contracting cancer at some future date, for each year of exposure at that level. The normal annual risk at age 30 is approximately 1 : 2000, falling to 1 : 400 for a heavy smoker. The average environmental radiation dose is about twice the ICRP limit, i.e. 2 mSv. The range quoted for the risk factor represents the ‘official’ view of ICRP (endorsed by the UK National Radiological Protection Board (NRPB)) of 1 : 100 per Sievert and those of the critics who would increase that to 1 : 10 per Sievert, with several prominent researchers arguing for a figure about half-way between the two extremes. Thus, environmental radiation burdens are clearly not without risk, but arguably, they could be described as ‘safe’ rather than ‘unsafe’. It may become clear that the word ‘safe’ has far more use as a political pacifier than in scientific debate. Unfortunately, media simplification often requires a ‘safety’ limit, above which something is ‘unsafe’.

Having considered individual dose limits, the societal impact must be considered. The sum of all the individual doses is known as the ‘collective dose’. For example, 200 million people in Northern Europe have regularly received a dose of 200 person-Sievert/annum from Sellafield discharges. This would be equivalent to 2–20 deaths for each year of exposure at this level. How is it then decided whether this level of exposure is acceptable?

According to ALARA, social and economic factors should be taken into account. This has been interpreted as requiring some form of cost–benefit analysis whereby the collective dose is ascribed a monetary value to represent the health costs to society.[7] As might be expected, various values have been used and proposed. NRPB recommended a figure of £3000 per person-Sievert for low doses (and a sliding scale when there is a significant proportion of higher doses), whereas the US Environmental Protection Agency have used $100 000 per person-Sievert. It is clear that annual 'detriment' has run to several million pounds on these criteria.

Thus far we have explored the impacts of pollution within the context of the numerical indices used by ICRP/NRPB and approved by government. We have seen that costings can be applied and that health effects are predicted within a framework of 'acceptable' damage following the ALARA principle. It ought to be clear from the preceding section that this apparently scientific process is replete with value judgements and politically expedient methodology, even before one considers any uncertainties in the modelling and predictions. It is perhaps less surprising in the light of this system that other foreign governments with ready access to scientific expertise have found disagreement with British assessments.[8]

NON-QUANTIFIABLE IMPACTS

That there are environmental impacts arising from low-level radioactive pollution which are not amenable to quantification was recognized by the ICRP in 1973. In Publication No 22, the Commission recommended that 'aesthetic and other human factors' be taken into account. Few national authorities have found an effective way of doing this. The NRPB have attempted consultations[7] and the Holliday committee recommended some form of social impact study for sea-dumping which led to the development of the Best Practicable Environmental Option study in 1985[9].

The problem is that the 'aesthetic' component of 'pure' food, and environmental 'quality' related to amenity, all rely on public perception and values which vary among different sectors of society. This social 'context' will clearly affect the interpretation of monitoring results, as well as the nature and extent of the monitoring exercise and presentation of its results. The scientists and regulators will be subject to their own perceptions which tend to discriminate toward that which is technical and quantifiable.

The existence of divergent 'world views' and the way it affects interpretation and response to technical information has been subject to study by the International Atomic Energy Agency in precisely this context. Unpublished work by Dutch social psychologists, Stallen and Meertens in 1976, took account of such divided perceptions of nuclear power risks, leaning heavily on two categories of response proposed by the philosopher Habermas: 'technocratic' and 'interactive'. The technocrat saw progress as material, outside of the Self and thus responded positively to any growth in complexity and control of natural resources; whereas the 'interactive' (for which we could substitute the 'environmentalist') saw progress as essentially internal, in terms of human values and harmony within the community. When the different respondents were presented with a set of agreed technical facts on nuclear power, both sets evaluated the technology according to their predisposition. A major conclusion of the study, of relevance to the IAEA's information programmes, was that the amount and type of information would do little to alter the evaluation of nuclear technology: technocrats would approve: environmentalists disapprove.

POLLUTION, DISCHARGE CONTROL AND THE BROADER ENVIRONMENTAL DEBATE

In the case study of Sellafield's radioactive discharges there are elements of relevance to the broader environmental debate on pollution control of chemical discharges or dumping in the environment. Although there are seldom any cross-references on the part of radiologists or radio-ecologists, the UK approach on radioactivity parallels that for non-radioactive substances in which Britain is at variance with the rest of the EEC (and to a certain extent the rest of the world). Britain is certainly the main antagonist with regard to new initiatives based upon the 'precautionary principle', put forward by West Germany in moves to protect the North Sea.[10] Briefly, Britain has argued that it was safer and more practicable to set discharge levels according to the capacity of the local environment to absorb the pollutant within the bounds of acceptability, rather than to strive for uniform emission standards. Thus the UK set Environmental Quality Objectives or Standards (EQO/EQS) for coastal waters and rivers and discharges are licensed according to their calculated effect on these EQOs. Thus, where a technology to control the discharge exists (presumably at additional expense) it need not be deployed if the overall EQOs are met. Within the rest of the EEC the Emission Standard approach produces a requirement on firms to update with the best available technology especially where pollutants are highly toxic, and tends to set the same standard such that no companies are unfairly discriminated against. European countries and industries have had long-running quarrels with Britain on the grounds that its standards were lax, thus creating unfair commercial advantage. In response, the British have argued that the coastal environment and rivers of the UK were fundamentally different (short, fast flowing rivers and rapid tidal flows) and allowed more flexibility. Clearly there is some logic in having tighter standards for a sluggish poorly flushed marine basin than one where dilution is rapid and effective.

The British approach, however, puts a much higher safety premium on reliable monitoring and interpretation of those results for animal and human ecology. The EQO must be set with reference to an adequate ecotoxicity data base and model of accumulations and future environmental changes, especially for persistent, bio-accumulative substances.

We have not time to explore this overall approach in more depth. Suffice it to say that the UK policy with regard to radioactive contaminants is no exception to the general policy, and indeed is a clear example of how the UK policy is prone to severe miscalculation on both scientific and public perception grounds, with expensive consequences (for a review of the Sellafield issue in an international context, see Taylor[11]).

THE CONCEPT OF LIMITING ENVIRONMENTAL CAPACITY AND THE HISTORY OF SELLAFIELD DISCHARGES

Reprocessing activities began at the Sellafield site in the early 1950s with the atomic weapons programme. In 1955, at the UN Conference on 'Peaceful Uses of Nuclear Energy', the international scientific community were first informed of a 'deliberate experiment' in Cumbria, which had taken the form of discharges of radioactive liquid effluent by pipeline into the Irish Sea.[12] These experimental discharges were intended to elucidate the fate in the environment of a cocktail of radioisotopes with widely differing chemistries. It is noteworthy, and we shall return to this in more detail, that other nuclear weapon states which were also developing reprocessing plant (USA, USSR and France), took a more cautious approach, particularly with regard to the long-lived plutonium isotope, and utilized maximal effluent treatment and recycle such that offsite liquid discharges were all-but eliminated. This contrasting approach, which was presumably known to British scientists, was not widely discussed until the present controversy began around 1975.

The small-scale reprocessing for the military programme was increased in 1964 by the construction of a second 'Magnox' plant (magnesium alloy casing for uranium metal fuel) to accommodate fuel from the growing programme of civil reactors. Discharges increased steadily in this period, but it was the early 1970s which saw the largest increases. The radioactivity is generally presented in two categories of discharge: alpha-emitters (e.g. isotopes of plutonium and americium, also known as actinides), and beta-gamma emitters (caesium-137 and strontium-90 being important examples). The actinide discharges arose largely from treatment of accumulated wastes from military operations, and the caesium primarily from a failure to foresee holdups in pond storage of the Magnox fuel, which corrodes rapidly in water, and the consequent lack of pond-water treatment systems which have only recently been commissioned.

The pattern of discharges can be seen in Figures 1 and 2 for both alpha and beta-gamma emitters. The major contribution from americium-241, plutonium isotopes and caesium-137 is highlighted. The full range of nuclides is not considered here, as caesium and plutonium/americium serve to illustrate the relation of radioecology to discharge policy. They represent two ends of a spectrum with regard to behaviour in the environment: caesium behaves conservatively in seawater (stays in the water column) and concentrates in biota, being evenly distributed in tissues, and emits a penetrating radiation. As a consequence of these properties it is readily monitored and its behaviour in the environment is reasonably predictable being largely determined by currents and the mixing properties of the surrounding seas. Moreover, with a half-life of 30 years (it will reach negligible concentrations after 300 years) it will not be subject to long-term bio/geochemical cycles. In contrast, americium and plutonium behave quite differently in the marine environment: they are in the main (95 per cent) removed rapidly from the water column by particle reac-

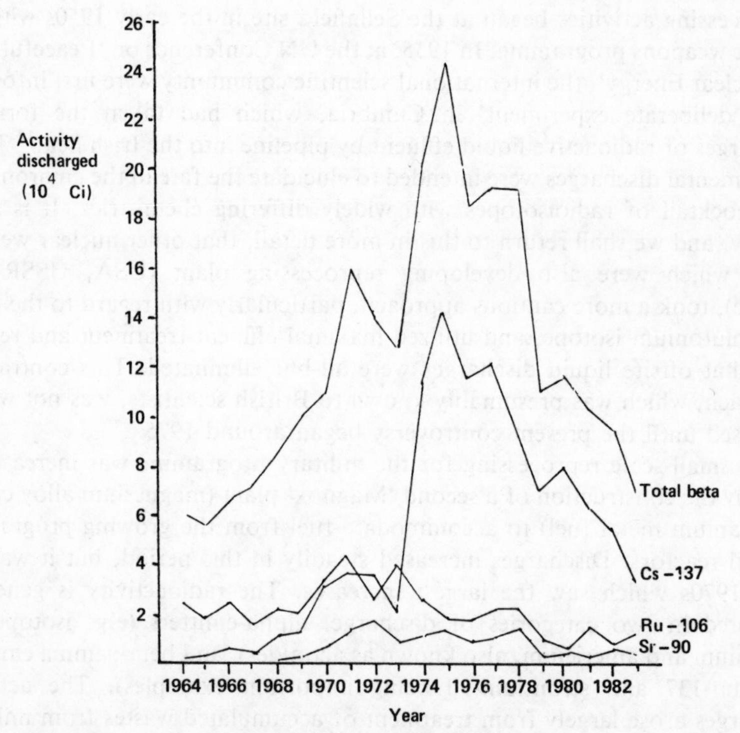

Figure 1 Annual discharge of selected beta/gamma-emitters from Windscale.

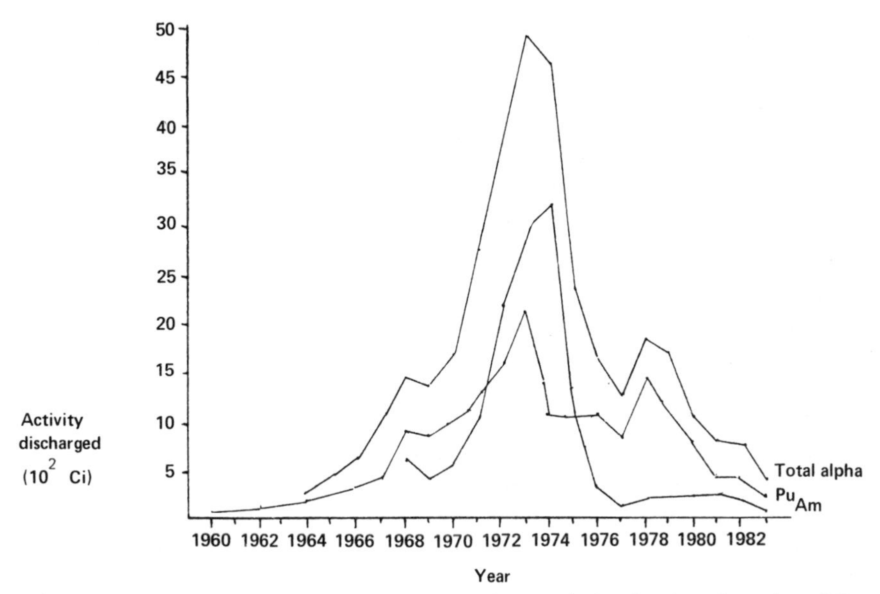

Figure 2 Annual discharge of alpha-emitters from Windscale. Pu: plutonium-238, -240, 238. Am: Am-241, not including decay from Pu-241.

tions and bound to sediments and thus not immediately taken into food chains (with the exception of coastal contamination of shell-fish filter-feeders); they are alpha-emitters, a short-range radiation that can only cause biological damage if incorporated either by ingestion or inhalation (plutonium-241 is a beta-emitter, but decays with a half-life of fourteen years to americium-241). The actinides have long half-lives (americium at 400 years, decaying to neptunium with $2\frac{1}{2}$ million years, and plutonium-239 with 24 000 years), and will thus be subject to long-term biological and geochemical cycles.

The radioecology of these discharges has been studied by the UK Ministry of Agriculture, Fisheries and Food, which has had responsibility for monitoring and is also a co-licensee with the Department of the Environment.[13] Detailed criticism of the monitoring and modelling first arose as a result of the Windscale Inquiry in 1977 and in submissions to the Department of the Environment by the Political Ecology Research Group (sponsored by Greenpeace) in 1982.[2,14] However, serious concern in the scientific community about the environmental effects first arose in 1975 when V. T. Bowen, a leading marine geochemist at the Woods Hole Oceanographic Institute (USA), and a specialist in radioecology of fall-out plutonium, warned the UK government of potential problems from the uniquely high levels of discharge (this led to a Granada TV documentary on the problem). The controversy heightened at the Windscale Inquiry when Professor E. P. Radford, a senior specialist in radiation epidemiology and Chairman of the US Academy of Sciences Committee on the Biological Effects of Ionizing Radiation (BEIR), voiced his concern for the health of the residents of the coastal region where plutonium levels were enhanced, (the Inquiry evidence is summarized in Stott and Taylor[2]).

From the discharge profile it can be seen that the peaks occurred before any major scientific or public concern was expressed. Although the major proportion of beta-gamma (caesium-137) resulted from inadvertent errors rather than conscious decisions, as admitted at the Inquiry, the actinide discharges were the result of conscious management decisions. It could be argued that even in the case of caesium-137, the lack of pond-water treatment in the event of failure, and decisions not to close the plant reflected a belief that highly contaminated effluent could be discharged if the occasion demanded.

Public concern began at the Inquiry where BNFL faced sharp criticism from the Isle of Man Local Government Board and Cumbrian and Lancashire Fisheries organizations. A commitment was made to reduce the discharges and as is evident, by 1983 they were one-tenth of the peak levels. The latest figures for 1985 show total beta-gamma at 587 TBq (15 000 Ci) compared with 2000 TBq (54 000 Ci) in 1983, and 6 TBq for alpha compared with 20 TBq in 1983.

However, as can be seen from Figures 3–5, caesium-137 and to a lesser extent plutonium-239 (5 per cent of the discharge is a soluble fraction), has dispersed widely. The caesium-137 is readily measurable in the Arctic seas off Greenland and plutonium levels off the west coast of Denmark are 100 per cent greater than those off the Baltic coast (the benchmark from weapons fall-out). This dilute and disperse philosophy leads to very low, but nevertheless measurable contamination of fish stocks throughout the Irish and North Seas and is much more pronounced in the local Cumbrian and North Lancashire fisheries. In the case of North Irish Sea fish and shell-fish, radiation levels in sea-food are still far in excess of that from the natural radioisotopes of potassium-40 (Figures 4 and 5), although the less well-known natural alpha-emitter polonium-210 must in many cases exceed the plutonium alpha contamination. The natural radioactivity of seawater has been elevated by the additional artificial nuclides by a measurable amount. Parts of the Irish Sea coast have a 100 per cent increase and small percentage increases are notable off Ireland and the western reaches of the North Sea.

Despite the reductions, brought about by expenditure of several hundred million pounds, the contamination of fish and coastal ecosystems is still the subject of intense political pressure on the UK government to bring about 'near zero' discharges by adopting the best available technology. In 1985 the Isle of Man Government called for the closure of the plant if this target could not be met. The Irish and Scandinavian governments have sought to pressure the UK government via the regional convention on land-based discharges known as the Paris Commission to cease the discharge practice. Greenpeace first drew attention to the health implications and lack of effective control in submissions to the Paris Commission in 1983, followed by detailed evidence on the availability of control technology.[15,16]

LEC and acceptable dose

As noted, the UK authorities and BNFL (then the UKAEA) took a conscious management decision in the 1950s to dilute and disperse substances which were

Figure 3 The spread of caesium-137 between 1973 and 1978 and plutonium levels relative to fall-out (1975). Caesium-137, pCi/l ------ 1973, – · – · – · – 1978 2× Plutonium relative to fall-out levels.

known to be highly toxic. This was in contrast to the precautionary approach taken elsewhere at that time. The plant was rebuilt in 1964, a time when the French also built a new reprocessing plant and when the USA upgraded theirs, and when 'virtual zero' discharge technology had been tried and tested at various inland plants worldwide (USA, USSR and France). In the 1960s discharges steadily increased. In the early 1970s MAFF promulgated Limiting Environmental Capacity (LEC) values for plutonium based upon the ICRP

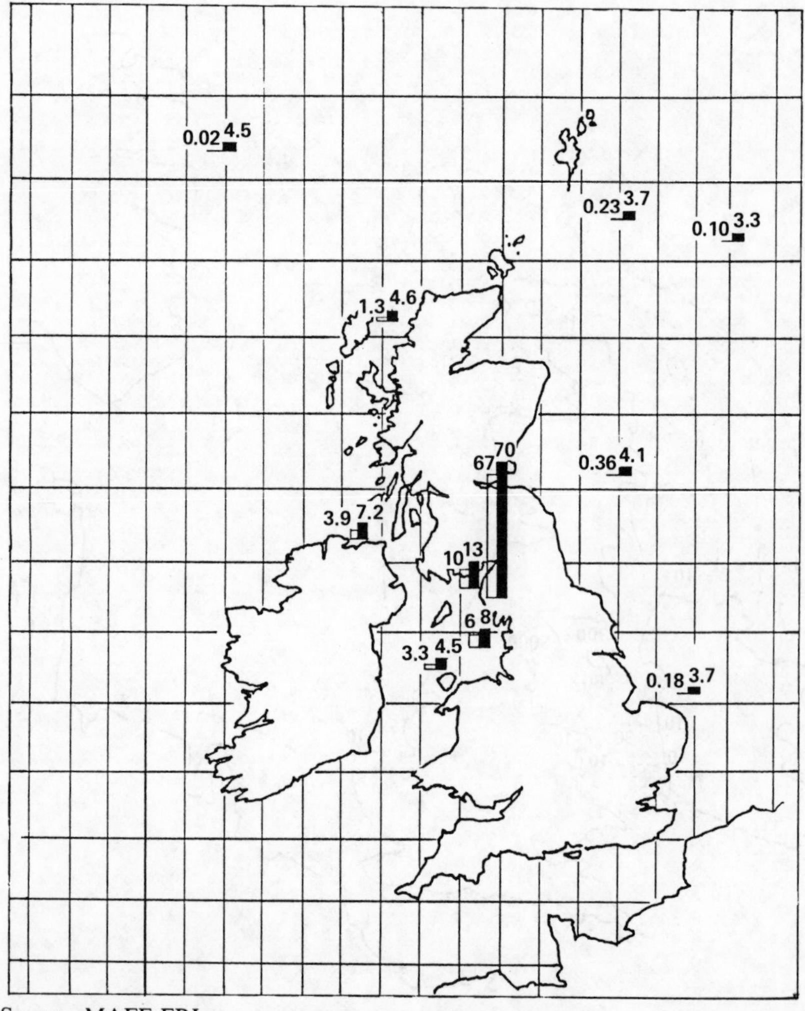

Source: MAFF FRL

Figure 4 Radioactivity recorded in fish around Britain, 1983. Gross beta (solid Bars) and caesium-137 (open bars). Units: pCi/g wet.

limits of 5 mSv for the most exposed members of the public (critical group concept). There were (and still are not) no standards for the protection of marine biota, and no standards for non-living environmental materials, other than those relating to human doses (i.e. no dose, no environmental impact). In the case of alpha-emitters that were apparently removed from the immediate food-chains to man, the LEC obviously appeared to be high, and was first calculated at 2664 TBq/yr (72 000 Ci/yr). A discharge of 148 TBq/yr was

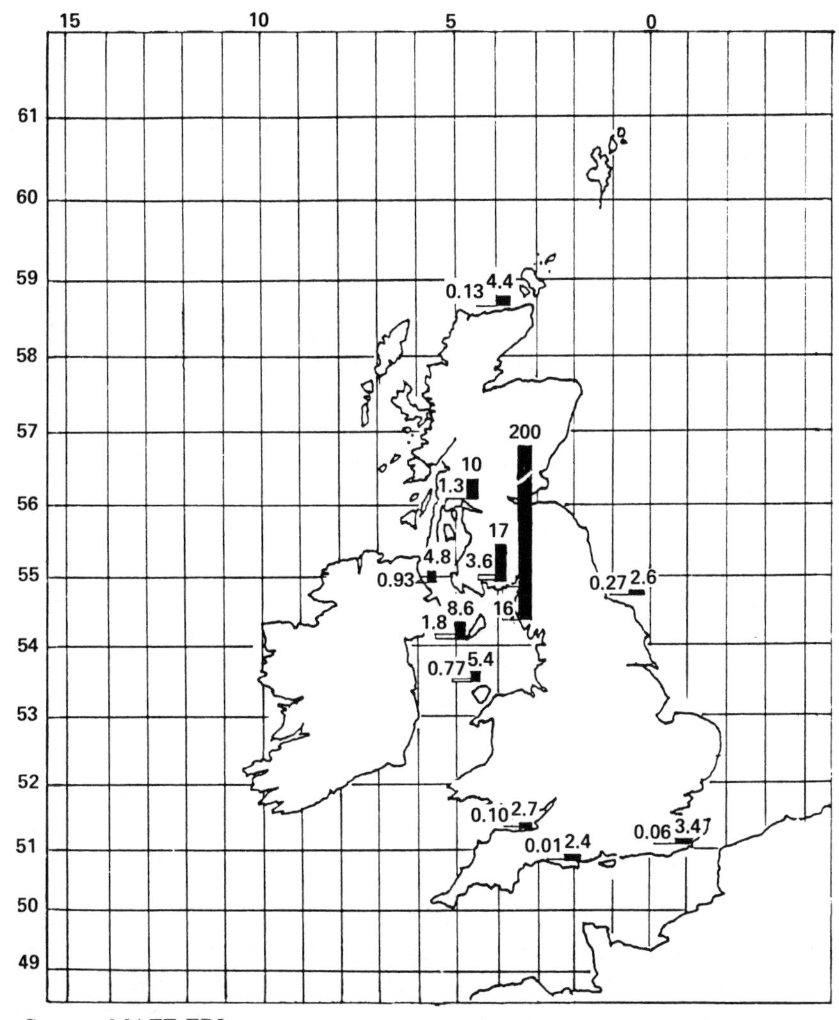

Source: MAFF FRL.

Figure 5 Radioactivity recorded in shellfish around Britain, 1978. Gross beta (solid bars) and caesium-137 (open bars). Units: pCi/g wet.

duly authorized to provide a suitable safety factor. In 1970 this figure was revised to 222 TBq to allow increases expected from the reclamation processes on military waste.

Likewise LEC's were calculated for the other isotopes on the basis of local consumption figures for sea-food. The caesium-137 discharge reached 70 per cent of the authorization in 1974 (MAFF Fisheries Research Laboratory reports) and shortly thereafter the critical group were receiving approximately 50 per cent of the ICRP 5 mSv dose limit. Figure 6 shows the maximum concentrations of nuclides in local sea-food and Table 1 the respective doses arising from the consumption model. Experimental verification of these dose levels by whole-body monitoring was not carried out until after the Windscale Inquiry and then not systematically for the critical group, but it is generally thought that the model over-estimates doses via the sea-food pathway, perhaps by a factor of 2.

It can readily be seen that the Limiting Environmental Capacity approach has several pitfalls:

(i) The environment may not be as predictable as the current state of scientific knowledge supposes. Thus the gradual build up of plutonium along the Cumbrian cost (Figure 7) is attributable to sea-to-land transfer of actinides.[17,18] These mechanisms were not anticipated by the regulatory authorities.
(ii) Even if everything follows according to the model, what may be an acceptable level of pollution in 1970, may not be acceptable in 1990 as public knowledge and expectations change, or new scientific data arises on the toxicity of the substance discharged.

The situation may be remedied quickly if the environmental levels are directly related to the annual discharge (although cutting the discharge may be expensive and controls take time to install—BNFL commissioned a pond-water treatment plant in 1985 at a cost of 200 million pounds eight years after criticism at the Windscale Inquiry). However, environmental levels may be subject to the cumulative load, especially if diagenic processes release the substance from sediment, or contaminated sediment is subject to erosion.

The dispersal of the apparently 'buried' alpha-emitters, something predicted by Bowen in 1975, is only now discernible from MAFF monitoring data. In Figure 6 it is apparent that shell-fish contamination peaked in 1982, some eight years after the americium discharge peak of 1974 and it is not clear what ecological factor caused this. From the columns of sediment contamination in Table 2 it can be seen that americium in Cumbrian sediment was little changed in 1983 to that in 1974 and that concentrations south of Morecambe Bay (Rock Ferry) had peaked in 1982 and were marginally greater in 1985 than in 1977 at 180 Bq/kg compared with 170. Levels in Conway mussels (North Wales coast) in 1985 were similar to 1980 and Wirral mussels showed little change in

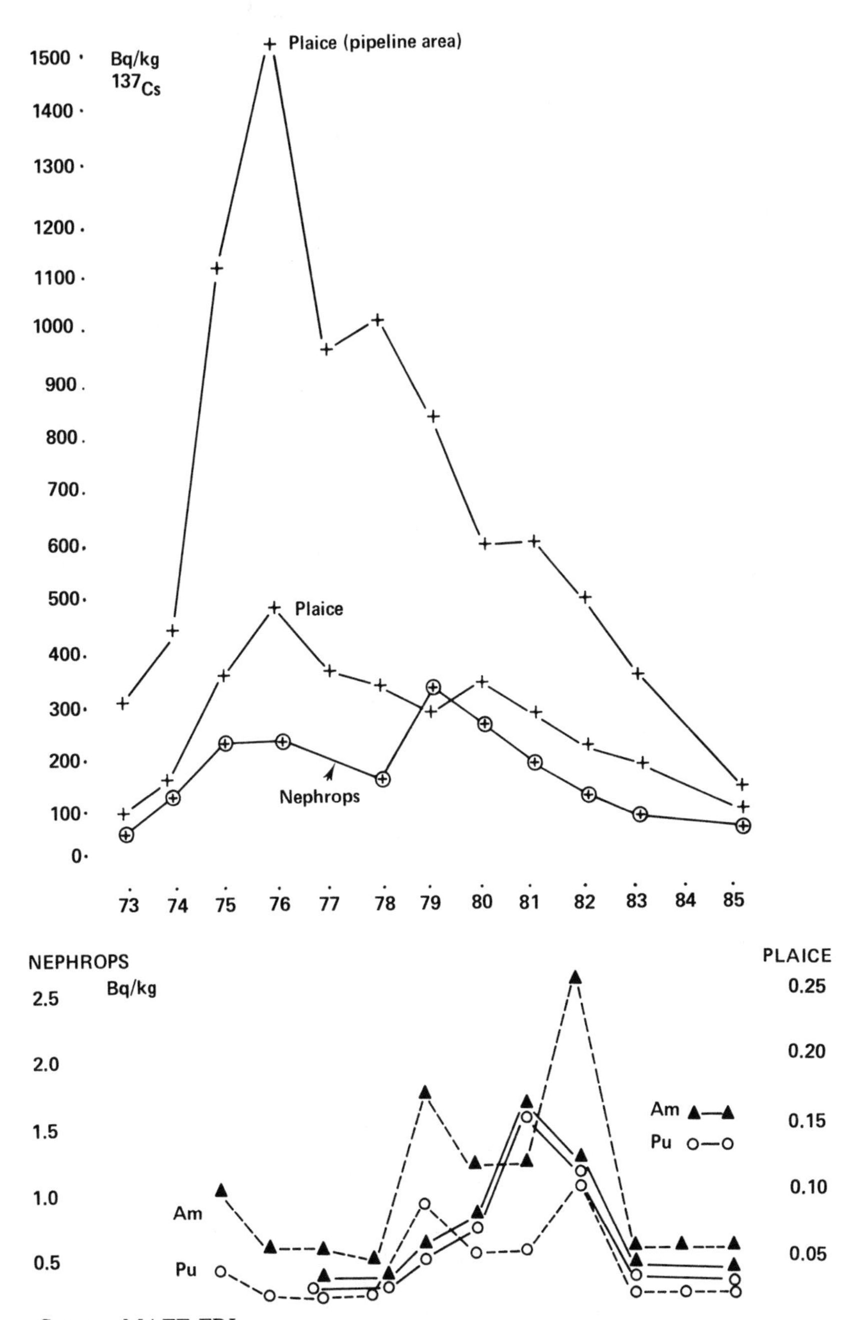

Source: MAFF FRL.
Figure 6 Concentrations of ^{137}Cs, ^{239}Pu, ^{241}Am in edible parts of fish and shellfish close to Windscale.

Table 1 Doses to various consumers and changes in consumption rates of fish and shell-fish from the Irish Sea

		mSv			Man-Sv	
	Pu	Max. L	Max. C	Av.	CD (UK)	CD (EEC)
1985	(a) × 5	0.7	0.03	0.02	30	50
1984	(a) × 5	0.8	0.3	0.02	70	100
1983	(a) × 5	2.25	0.5	0.03	70	110
1982	(a) × 5	2.70	0.6	0.04	90	110
1981	(a) × 5	3.45	0.9	0.06	130	150
1980	(b) × 5	1.95	0.9	0.05	100	140
1979	(c)	1.01	0.7	0.05	130	170
1978	(c)	1.27	0.7	0.05	128	107
1977	(c)	1.50	0.6	0.06	89	80
1976	(d)	2.20	0.8	0.12	140	120
1975	(d)	1.70	0.6	0.04	83	57
1974	(d)	0.70	0.1	<0.04	43	30
1973	(d)	0.15	<0.1	<0.04	15	15
1972						

	Consumption (g/day)*		
	Max. L	Max C	Av.
(a)	100 18 45	360 70 50	40
(b)	100 18 18	360 70 50	40
(c)	170 15 6	360 70 50	40
(d)	224 41 —	290 70 45	40

* For (a), (b), (c) and (d), where there are three entries in a column, the first entry is the consumption of fish, the second crustaceans, and the third molluscs.

Max.L is the maximum local consumer (actually an average figure for the critical group). Max.C is the maximum consumer associated with the commercial fisheries at Whitehaven. Av. is the average domestic consumer. CD = collective dose. From 1980, Pu uptake factor increased × 5.

five years (data prior to 1980 is not readily available). Despite the paucity of monitoring data, these figures support a gradual movement of alpha-contaminated sediment south from the Cumbrian coast. The fact that these levels are still very low and of 'negligible' contribution to the natural environmental dose may do little to appease public opinion regarding the loss of 'clean' seas and uncontaminated food.

Remaining with the issue of individual doses, the situation in Cumbria itself has led to widespread public concern, despite assurances from the scientific

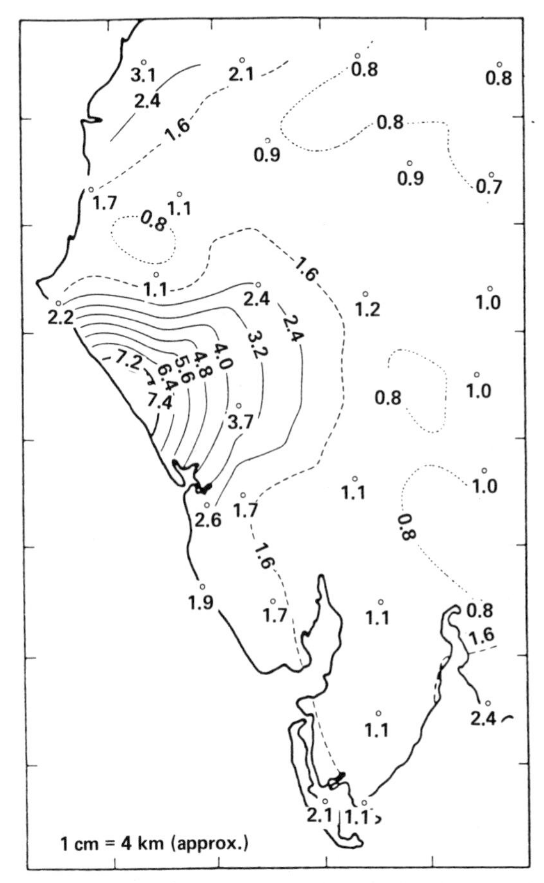

Figure 7 Deposition of plutonium-239 + 240 relative to estimated fallout in Cumbria, 1978. From Ref. 17. Notes: (i) data are for grassland sites, (ii) contour interval is 0.8, (iii) samples are from top 15 cm of soil. Eakins *et al.*[18] estimate that a total of 1–2 Ci has been deposited in the 5 × 40 km coastal strip. This represents a 100 per cent increase in fallout levels, estimated at 1 Ci for the region.

Table 2 Americium-241 concentrations in Bq/kg for sediment and shell-fish in the Irish Sea

	1985	1984	1983	1982	1981	1980	1979	1978	1977	1976	1975	1974	1973
Newbiggin	1900		2700	2500	2300	2800	2600	2900	3300		(2738)		
Walnty I.	330		450	570	780	800	750	—	—				
Rock Ferry	180		150	240	240	150	200	160	170				
Conway mussels	0.31		0.61	0.18	0.36	0.38	—						
Wirral mussels	5.4		4.9	1.8	4.0	5.3	—						
Morecambe cockles	7.5		9.1	6.0	9.1	8.0	7.1						
Ravenglass cockles	76.0		75.0	93.0	—	—	—						
Sellafield mussels (St Bees)	46.0		37.0	62.0	78.0	81.0	99.0	300	340	340	—	185	

professions that modelled doses are still well within, if not now 'acceptable' limits (finally admitted), then at least levels which could not cause an observable health impact. The problem with this, is that an observable excess of a radiation-linked disease (childhood leukaemia) has now been proven for the Cumbrian coastal strip, the precise area of elevated plutonium levels.

The plutonium dose to the most avid sea-food consumers has indeed risen, despite the falling discharge levels. Figure 8 demonstrates this. In 1978 the maximum consumer dose was dominated by the caesium isotope and was thus relatively predictable and amenable to control by the annual discharge. However, a number of changes had to be made to the model as a result of research (much of which arose from PERG's criticisms at the Inquiry in 1977, for example, of the absence of transfer coefficients for organically bound plutonium). The model was changed to accommodate a new gut-transfer factor (500 per cent increase) and also a shift toward greater shell-fish consumption. The result can be seen in the 1981 column, where actinides account for 70 per cent of the dose. It should be noted that the rapid reduction between 1983 and 1985 to 0.8 mSv is the result not so much of lower environmental levels, but drastically reduced consumption following the 25-mile public exclusion of the coastal strip following an accidental spillage in November 1983. It is perhaps fortunate for BNFL that the NRPB should have settled upon a 1 mSv control limit at a time when the critical group had altered their consumption habits as a result of the plant's activities. However, the 1 mSv limit is still four times the US EPA limit at which an installation would have to close down.

This latter instance is an example of how the marine environment itself is not protected by a physical EQS, but only in so far as it relates to doses to

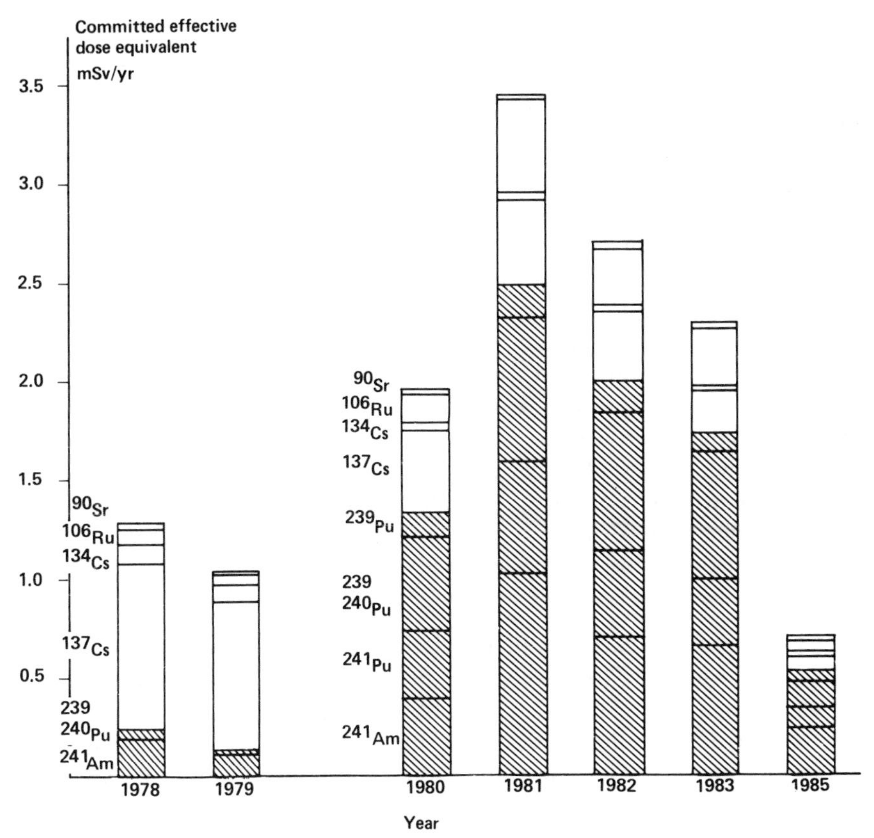

Source: MAFF FRL.
Figure 8 Individual radiation exposure due to consumption of Irish Sea fish and shellfish: consumers in the local fishing community (critical group).

man. Although it is unlikely that significant biological damage could occur to a coastal ecosystem within the present regulatory framework, it cannot be ruled out for local populations of carnivores such as otter, seal and possibly some sea-birds. It should also be apparent that public reaction to low levels of contamination is not taken into account by the present dose-standard.

Against this background of scientific uncertainties and the difficulties of modelling, with reputable scientists involved in public disagreement over the interpretation of monitoring results, it is perhaps not so 'irrational' for a lay public to demand a 'precautionary' approach, particularly if containment is available at reasonable cost. The issue is compounded by the results of the Black Inquiry, which had as its initial remit to establish whether the excesses of leukaemia were real.[19] Having established that they were, but that there was no proven radiological cause, many recommendations were made for tightening up the monitoring and validating the models. The 'qualified assurances' of the Black Report did little to allay local fears, as evinced by concern from district councils, the Isle of Man and the Irish Government during 1986 (see Isle of Man,[20] Paris Commission[8]).

The question of whether the observed health effects are caused by the environment levels is regarded as open, although most regulatory professionals regard it as highly improbable. It is largely a question of whether the accepted areas of uncertainty are sufficient to account for the kinds of doses required. At this stage, speculation may help delineate further research directions.

At first sight it would appear that the maximum consumers (who are not necessarily directly linked to the children who died) were getting no more than twice the normal background exposure to ionizing radiation. There is nothing biologically unusual in this—it being a regular consequence of differing regional geologies. However, the doses are expressed by MAFF as 'dose equivalents' which have already used various weighting factors to allow for the organ exposed (alpha-emitters do not expose the whole body) and the biological effectiveness of the type of radiation. Now that the critical group receives most of its dose from alpha-radiation, it is pertinent to examine the following:

(i) The naturally occurring alpha-radiation regime for sensitive sites such as the bone marrow (relatively well protected from environmental alpha-emitters) is in the region of 0.2 mSv. In adults the bone-marrow dose from artificial radionuclides may be fifty times the normal alpha dose.
(ii) Particular uncertainty surrounds the bone-marrow dose in foetus and infants.
(iii) The transfer factors for suckling infants are not known other than by comparison with rodent data, which for plutonium show values 100 times the normal gut transfer.
(iv) The Relative Biological Effectiveness (RBE) of alpha-radiation compared with gamma-radiation in carcinogenesis is well known and estimated at twenty times, but this is derived from animal data and can only be discovered from epidemiological studies: there are none for the human infant.
(v) There is still controversy over the dosimetry arising from the exact distribution of plutonium in the bone and marrow forming areas. Given these uncertainties, controversy is bound to continue as to the likelihood of a causal link between the discharges and local cases of childhood leukaemia.

The case remains unproven either way; but one point should be reiterated that was made at the Windscale Inquiry in 1977, and also by the Royal Commission in 1976, and again to the Department of the Environment in 1984 by PERG, and it is that co-ordination of research between MAFF, the UKAEA, the NRPB and the DOE has been historically very poor, each having had responsibility for certain areas of research and none having an overview. Theoretically, DOE and NRPB took on a systems approach, but in 1984 the DOE were still resistant to the idea of subjecting the radiobiology to the kind of searching outside review to which they had subjected the radioecology. It is in this area that most uncertainty lies. The NRPB research that underlay the conclusions of the Black Report is available.[21]

THE USE AND ABUSE OF ALARA

Attention has been drawn already to the role of ALARA in the dose limitation system. The ICRP have stated that it is ALARA which is the most restrictive of the three principles, and the UK authorities have cited ALARA in defence of its upper limit standards (5 mSv or 1 mSv) which have been out of line with limits set by most other nuclear power states for public exposure as a result of discharges—these range from the US EPA limit of 0.25 mSv to the German 0.3 and Scandinavian 0.6. In the EPA case, installations still have to justify their doses with an ALARA-type consultation process and most achieve well below the limit. The UK have argued that their ALARA target for Sellafield is around 0.5 mSv, above which they pressure BNFL to install controls, as, indeed, has been the case. One is entitled to ask, however, whether ALARA is an effective system.

In the case of Sellafield, PERG research[11,16] has shown that:

(a) 'Virtual zero' discharge technology was available in the late 1950s (France, USSR and USA).
(b) The cost at the design stage for recycle plant was not significantly different to BNFL's wash-through plant.
(c) Even following criticism during the 1970s and submission to the Windscale Inquiry of detailed engineering specification from German recycle technology (for Thermal Oxide fuel), the British did not proffer a recycle plant for the new THORP facilities, and have only now agreed, after the intervention of the Paris Commission and foreign governments that THORP will be virtual-zero.

Tables 3 and 4 show how UK reprocessing plant has performed compared with overseas installations. On both the discharge and the dose standard, both UK plants are in a league of their own, reflecting totally different management philosophy to the environment.

The US data show that individual exposures of less than 0.01 mSv via ingestion (riverine ecosystem) were achievable in the early 1970s. The West German reprocessing project expects to meet a similar target. In 1978, UK experts reported to the Norwegian Royal Commission[22] that 0.14 mSv was a reasonable target for a model plant. Windscale doses have been 1000 times

Table 3 Relative performance of major reprocessing plants worldwide in terms of Curies discharges per tonne of heavy metal reprocessed

	Beta	Alpha	te	f	teE	beta/teE	alpha/teE
Dounreay 1964–74	100 000	400	7	15	105	1000	4
Windscale, 1972–82	2 000 0 00	35 000	10 000	1	10 000	200	3.5
Cap le Hague 1968–78	200 000	100	8000	1.2	10 000	20	0.01
WAK, Germany Total	100	1	200	8	1600	0.05	0.0006
Hanford, USA Total	400	40	10 000?	1	10 000	0.004	0.004

f: a factor derived from different burn-up categories of fuel which affects the radioactivity per tonne. teE: an equivalent representing the radioactivity throughput.

what is considered reasonably achievable elsewhere, and up to twelve times the US EPA regulatory limit for such operations.

When asked to provide the detailed cost–benefit analyses that are recommended by the ICRP for ALARA implementation, BNFL (and the authorizing departments) have responded that only broad judgements have been made, but that prospective health detriment has been taken into consideration. Early MAFF figures for the cost of a person-Sv of detriment were £5000 compared with the US EPA recommendations of $100 000. Recent NRPB recommendations are that a person-Sv arising from large numbers of exceedingly small doses should be costed at £3000.

Table 4 Comparison of Windscale with other reprocessing plants in terms of maximum annual discharge and maximum annual dose off-site from liquid discharges.

	Liquid discharges (Ci)		Maximum annual dose off-site (mrem/year)
	Alpha	Beta-gamma	
Hanford, USA			
1972*	4	500	0.5
1975 (EPA ruling)†	0.13	5	0.5
1981 (EPA ruling)‡	'0'	'0'	'0'
Marcoule 1967–78	0.4		
West Germany 'Project'	0.4	0.4	0.5
Windscale 1970s max.	4800	250 000	350

* Last year of operation. †For refurbishment. ‡For new plant

Table 1 provides details of the 'collective dose' to the UK and to other European countries. From this can be calculated both the monetary cost of the health detriment and the numbers of expected health effects such as cancers and hereditary defects (between 1–10 per 100 person-Sv).

	UK	*EEC*
Cost		
NRPB	$2 664 000	$3 717 000
EPA	$88 800 00	$123 900 000
Cancers	16–160	24–240

It is noteworthy that the US costings were based on more than simplistic actuarial accounting of the cost of cancer deaths, and included environmental quality detriment. It can be seen that the actual expenditure eventually justified for the control of this pollution (about £200 million) is the same order of magnitude as the $200 million detriment that would justify the expense were EPA guidelines to have been adopted.

PIPELINE DISPOSALS AND ACCIDENTAL SPILLAGE

The BNFL plant has been plagued by small spills of radioactive wastes through the marine disposal pipeline. The most spectacular occurred in 1983 when Greenpeace divers investigating the pipeline were contaminated, and as the slick of solvents washed onto the shore, 25 miles of beach were closed to the public for several months[23]. An estimated 6000 curies of ruthenium-106 (well within the annual allowance) was involved. There was widespread public alarm and reports of economic losses in fisheries and local hotels. There have been many lesser spills.[24,25]

The prospect of a major spill, following a loss of containment accident was alluded to by the Royal Commission on the Environment,[26] when it estimated a potential 50 million curies could be released (fifty times the amount routinely discharged over the lifetime of the plant). The RCEP only considered the cancer impact and that in no great detail, concluding that it would give rise to 'a few' unless fishing were restricted. The Commission did not consider the potentially huge economic losses that could occur due to public reaction: the Irish Sea Fisheries would probably be closed and the North Sea severely affected—even if levels remained below 'official' action levels. A marine 'Chernobyl' could lead to billions of pounds in damage. These points were made to the Dounreay Inquiry by the Scottish Fishermen's Federation and the Sea Fish Industry Authority in an attempt to persuade the designers against a sea disposal pipeline.[27,28]

The existence of a sea disposal pipeline leaves management with the ready option of discharging any on-site accumulations which might prove hazardous to their workforce. In an emergency, the sea disposal route might be the lesser of two evils. Furthermore when pipelines exist within a design concept, there is less incentive to design other systems for containing spills safely. It is noteworthy that the West German WAK plant at Karlsruhe and the Marcoule plant in France, on the Rhine and Rhône respectively, have had no recorded major spills that have threatened health or amenity. These are essentially closed-system, recycle plant using technology available since the 1950s and early 1960s.

THE PRECAUTIONARY APPROACH

In this chapter we have shown that the UK policy of dilute and disperse, Limiting Environmental Capacity (or Assimilative Capacity) and the ALARA principle have not prevented widespread pollution. Indeed, that pollution has necessitated expensive retrofitting to a plant that even when all modifications are complete (by 1990) will still operate at least ten times less efficiently than a state-of-the-art reprocessing plant. There have been unaccounted economic losses, prospective health damage and a widespread perception of a loss of environmental quality, alarm and unease, with additional loss of confidence in regulatory bodies and scientific research. All this could have been avoided by a precautionary approach as applied by the US, USSR and France at the inception of reprocessing. There is a strong case for indicting the UK for failing to effectively apply its own criteria of ALARA, and thus for not abiding by international agreements.

Furthermore, UK monitoring and models failed to predict the return of plutonium to land, and there are still significant uncertainties with regard to the sea-food pathway and the critical group, as evinced by the Black Inquiry. Yet, it has been regularly stated that more is known of the ecology of radioactivity in the environment than any other pollutant.

It is concluded that the Sellafield discharges represent the British approach *par excellence*, and that it has signally failed to protect the environment. If this can happen with the most intensively monitored and researched pollutant (i.e. radioactivity), then the UK approach does not augur well for other contaminants for which it is currently reproached (mercury, cadmium, lead and sewage among others).

At the North Sea Conference in October, 1984, the German delegation led an appeal to Britain, as the odd-man-out, to adopt the precautionary approach.[29] Essentially, if a substance is known to be toxic, there is a presumption against discharge, rather than waiting until research and monitoring of its effects following dispersal proves its ecotoxicity one way or another (for discussions see: Dethlefsen,[10] Krom[30] and Sperling[31]).

With added pressure for BNFL to reduce discharges to virtual zero (effectively the ageing Magnox plant would have to close as it cannot be economically converted to a recycle plant), the long debate is not over. Whatever the final result at the Sellafield site, the history of the Windscale discharges should prove an object lesson in the assumption by marine scientists that they can sufficiently predict the environmental consequences of releasing toxic substances which may affect future generations and who may expect higher quality standards than those left over from a time of industrial expansion and optimism regarding all things technical.

REFERENCES

1. Hon. Justice Parker, *The Windscale Inquiry*, HMSO, London, 1978.
2. M. Stott and P. J. Taylor, *The Nuclear Controversy: a Guide to the Issues at the Windscale Inquiry*. London: Town and Country Planning Association; Oxford: Political Ecology Research Group, 1980.
3. Royal Society, *Risk Assessment: A Study Group Report*. The Royal Society, London, 1983.
4. F. Holliday, *The Independent Review of the Dumping of Radioactive Wastes in the North East Atlantic*. HMSO, London, 1984.
5. P. J. Taylor, The disposal of nuclear waste in the deep ocean. *Research Report RR-15*. Political Ecology Research Group, Oxford, 1985.
6. R. J. Kayes and P. J. Taylor, Health Risks of Nuclear and Coal Fuel Cycles in Electricity Generation: a critical review of comparative assessments for the United Kingdom. *Research Report RR-13*, Political Ecology Research Group, Oxford, 1984.
7. National Radiological Protection Board, *The Application of Cost–benefit Analysis to the Radiological Protection of the Public: a Consultative Document*. NRPB, Didcot, Oxford. HMSO, 1980.
8. Paris Commission, *Recommendations of the Technical Working Group to the 8th Meeting*, Madrid, June 1986 and *submission of the Irish delegation to the 8th meeting*, on the subject of best available technology for the elimination of radioactive discharges. Paris Commission, London, 1986.
9. Department of the Environment, *Assessment of Best Practicable Environmental Options for Management of Low- and Intermediate Level Radioactive Wastes*. HMSO, London, 1985.
10. V. Dethlefsen, Marine pollution mismanagement: towards the precautionary concept. *Marine Pollution Bulletin*, **17**, 2, 54–7, Pergamon, Oxford, 1986.
11. P. J. Taylor, 'Radionuclides in Cumbria: environmental issues in the international context', in *Pollution in Cumbria*, pp. 47–54, Institute of Terrestrial Ecology, Merlewood, Cumbria, 1986.
12. H. J. Dunster, (1956) The discharge of radioactive waste products into the Irish Sea. Part 2: the preliminary estimation of the safe daily discharge of radioactive effluent. *Proc. Int. Conf. on Peaceful Uses of Atomic Energy*. 1st Geneva, 1955, vol. 9, 712–715 (1956).
13. G. J. Hunt and D. F. Jefferies, 'Collective and individual radiation exposures from discharges of radioactive wastes to the Irish Sea', in *Proc. Int. Symp. on the*

Impacts of Radionuclide Releases into the Marine Environment, 534–570. Vienna: International Atomic Energy Agency, 1980.

14. P. J. Taylor, The impact of nuclear waste disposals to the marine environment. *Research Report RR-8*. Political Ecology Research Group, Oxford, 1982.
15. P. J. Taylor, The discharge of radioactive wastes into coastal waters and its subsequent dispersal: a case of transfrontier pollution. Paper presented to the Paris Commission, Berlin, 1983. (*PERG TM-01-83*), Political Ecology Research Group, Oxford, 1983.
16. P. J. Taylor, The control of radioactive discharges from nuclear fuel reprocessing plant: the best available technology. Paper presented to the Paris Commission, Dublin, 1984. (*PERG TM-03-84*) Political Ecology Research Group, Oxford, 1984.
17. P. A. Cawse, Studies of Environmental Radioactivity in Cumbria, Part 4: Cs-137 and plutonium in soils of Cumbria and the Isle of Man. *AERE R-9851*. HMSO, London, 1980.
18. J. D. Eakins *et al.* Studies of Environmental Radioactivity in Cumbria, Part 2: Radionuclides deposits in soil in the coastal region of Cumbria. *AERE R-9873*. HMSO, London, 1981.
19. Independent Advisory Group, *Investigation of the Possible Increased Incidence of Cancer in West Cumbria* (Black Inquiry), HMSO, London, 1984.
20. Isle of Man Local Government Board, *Report to Tynwald*, British Nuclear Fuels Ltd, Sellafield Operations, Douglas, Isle of Man, 1984.
21. G. S. Linsley, *et al.* An assessment of the radiation exposures of members of the public in West Cumbria as a result of the discharges from BNFL, Sellafield. *National Radiological Protection Board, R-170*, Didcot, Oxford, 1984.
22. Norges Offentlige Utredninger, *Nuclear Power & Safety*. (English edition), Government Printer, Oslo, 1978.
23. HSE (Health and Safety Executive), *The Contamination of the Beach Incident at British Nuclear Fuels Ltd, Sellafield, November 1983*. HSE, London, 1984.
24. HSE (Health and Safety Executive), *The Leakage of Radioactive Liquor into the Grounds, Windscale, 15 March 1979*. HSE, London, 1980.
25. HSE (Health and Safety Executive), *Safety Audit of BNFL, Sellafield 1986*. HSE. HMSO, London, 1986.
26. Royal Comission on Environmental Pollution, 1976, 6th Report, *Nuclear Power and the Environment*.
27. P. J. Taylor, The implications for Scottish and North Sea fisheries of the projected development of the European Demonstration Fast Reactor Fuel Reprocessing Plant at Dounreay, Caithness. *Research Report RR-16*. Political Ecology Research Group, Oxford, 1986.
28. N. McKellar, Economic significance of the fishing industry associated with the North Sea. *Submissions of the Scottish Fishermen's Federation, Aberdeen and the Sea Fish Industry Authority to the Dounreay Inquiry, 1986*. Department of the Environment, London, 1986.
29. K. R. Sperling, Prevention of Sea Pollution by Dumping at Sea—A report on the Scientific Working Group Related to the Oslo and London Dumping Convention. In German. *Vom Wasser*, Vol. 62, pp. 52–61. Verlag Chemie, Weinheim, W. Germany, 1984.
30. M. D. Krom, An evaluation of the concept of assimilative capacity as applied to marine waters. *Ambio*, **15**, 4, 208–214. Pergamon, Oxford, 1986.
31. K. R. Sperling, Protection of the North Sea, Balance and Prospects. *Mitteilungen der Deutschen Gesellschaft fur Meeresforchung*, Vol. 4, 3–8. English

translation available from the author, BAH-Labor Sulldorf, Hamburg, Germany, 1984.
32. HSE (Health and Safety Executive), *Report on the Silo Leak at Windscale*. HSE. London, 1980.
33. International Commission on Radiological Protection (ICRP), Implication of the Commission recommendations that doses be kept As Low As Readily Achievable. *Publication No. 22*, ICRP, Pergamon, Oxford, 1973.
34. Ministry of Agriculture Fisheries and Food (MAFF) Radioactivity in surface and coastal waters of the British Isles, 1966–1985 (*Aquatic Environment Monitoring Reports*), Lowestoft, UK, 1967–1986.
35. Isle of Man Local Government Board, *Report to Tynwald*, British Nuclear Fuels Ltd, Sellafield Operations, Douglas, Isle of Man, 1986.

Radiation and Health
Edited by R. Russell Jones and R. Southwood

4

The Role of Predictive Modelling: Social and Scientific Problems of Radiation Risk Assessment

David Crouch
Science Policy Research Unit, University of Sussex, Falmer, Brighton, UK.

ABSTRACT

This chapter starts from the premise that predictive modelling of radiation hazards, as it is currently practised, fails to reflect accurately the problematic status of the science of radiation toxicology. The chapter surveys the scientific problems facing the construction of robust models, arguing that uncertainties are such as to confine 'prediction' to order-of-magnitude analysis which can provide little more than a range of plausible risk estimates. It examines the origins of controversy over these issues and briefly analyses the former in terms of their cognitive and institutional dimensions. These problems are discussed in relation to recent concern over the raised incidence of child leukaemia around the Sellafield nuclear fuel reprocessing plant in the north of England.

If predictive modelling is to play a constructive role in debates over nuclear technology, the chapter argues that decision-makers and risk analysts should have greater regard for the extent of social and scientific dissensus surrounding the issue of radiation hazards. It concludes that this would best be achieved within the context of a less technocratic approach to risk assessment, and the establishment of formal mechanisms for public and worker involvement in the regulation of nuclear power.

INTRODUCTION

The predictive modelling of population doses from environmental radiation is a complex business. While I believe that modelling is scientifically worthwhile, however, I share an opinion that the sciences of radioecology, radiobiology and radiation dosimetry represent an inadequate basis for accurate assessment of environmental risks.[1-6] I consider that there is a need for greater philosophical reflection upon the epistemological status of these sciences. This I hope to demonstrate by means of a pertinent example: the controversy over the raised incidence of child leukaemia in the vicinity of the Sellafield nuclear fuel reprocessing plant in Cumbria on the north-west coast of England.[7] In the second part of this chapter, I shall discuss some more general questions regarding attitudes to risk assessment.

SCIENTIFIC PROBLEMS OF RISK ASSESSMENT

Attention has focused on the finding that deaths from leukaemia in childhood in the village of Seascale—just to the south of the plant—are approximately ten times the regional average.[8] In order to assist interpretation of this statistic, the National Radiological Protection Board (NRPB) calculated the number of childhood leukaemia deaths in Seascale that would have been expected following exposure to radioactive effluent from the plant.[9] This long and detailed risk assessment arrived at a figure of 0.01 leukaemia deaths among the Seascale child population since operations began at Sellafield in 1950; this figure has recently been amended to 0.016.[10] Thus the difference between observation and calculation is a factor of about 250. In reply to criticism of the risk assessment, the Director of the NRPB, Dr John Dunster, has stated that the Board has 'sought diligently for plausible sources of error of this magnitude and failed to find them'.[11]

In any risk assessment of this kind there is, I shall argue, very considerable scope for misjudgement and error. I shall concentrate on the uncertainties surrounding the actinides—the long-lived alpha-emitters plutonium and americium. I shall do this for two principal reasons: first, unlike beta-emitters, alpha-emitting radionuclides cannot be measured directly in the human body, thereby introducing an additional source of uncertainty; and secondly, in the case of beta and gamma emissions, considerations of microdosimetry are unimportant—for alpha-radiation they may be paramount. According to the NRPB's risk assessment, 12 per cent of the total Sellafield dose to the red bone marrow is high LET; thus for the actinides to have been responsible for the Seascale leukaemias there would have to be an error in excess of three orders of magnitude.

The outline structure of the NRPB's calculation is shown in Figure 1. The Board modelled exposure, metabolism and dosimetry to estimate the resultant dose to radiosensitive tissue in an average child. A risk factor was then employed to calculate the risk of death from leukaemia, and this was then summed over the entire child population. Finally, a sensitivity analysis was performed to investigate the significance of uncertainties in the models and parameters. Note that the four stages in the calculation fit together rather like Russian dolls: an uncertainty or error in any one stage requires modification

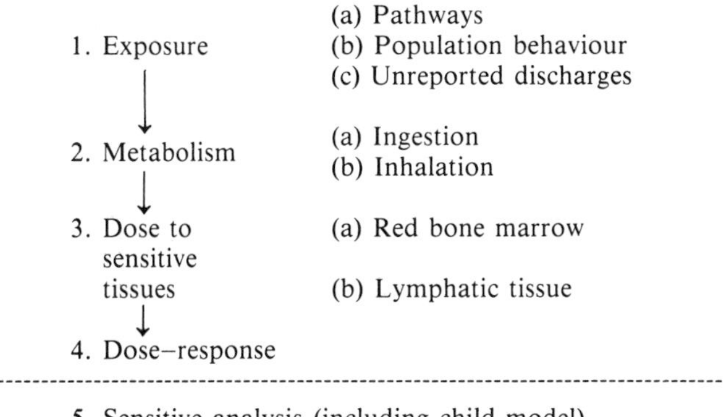

Figure 1 Structure of Seascale Risk Assessment Calculation.[9]

to the whole, in order that it neither rattles nor splits. Errors *between* stages are thus multiplicative; errors *within* stages are roughly additive.

I shall now go through the uncertainties in each stage of the calculation. In this I have been anticipated by Dr M.C. Thorne of ICRP, who has estimated that the actinide dose to sensitive tissue could be 'talked up' by two orders of magnitude.[12] I shall draw upon, and extend, his analysis.

Exposure

Pathways

Monitoring data at Sellafield are sparse: the first year for which comprehensive data on environmental concentrations are available is 1978.[9] Thus the concentrations of most radionuclides prior to the mid-1970s had to be estimated from reported levels of annual effluent discharges from the plant. For caesium isotopes and other radionuclides that remain in the water phase, concentrations in sea-water are closely related to current discharges. The actinides, however, bind closely to sediments. As a result, concentrations in marine foods and sediments may be significantly affected by the discharges of years gone by. The most recent review by Hunt reports that concentrations of Pu and Am may be related to discharges over a period of anywhere between one and eight years.[13]

Further uncertainties surround the resuspension of contaminated sediments in sea-spray. The actinide content of sea-spray over contaminated sediments may be many thousand times that of contaminated water free of sediment, pointing to some significant concentration mechanism in the surf.[14] Hamilton and Clarke recently suggested that a climatic event such as a severe storm might mobilize most of the actinide content of coastal sediments.[15] The potential significance of this exposure pathway was recognized only recently, and the NRPB were the first to have constructed a detailed model of the process.

Population behaviour

Crucial to an estimate of exposure is the behaviour of the exposed population. There were no measurement data on the habits of Seascale children, and in assessing average intakes of contaminated food and residence times on the beach (where intakes of contaminated sand, water, or sediment might have occurred) the NRPB relied largely on its own judgement.[16] A small proportion of children also deliberately ingest sand and mud. These considerations relating to the behaviour of Seascale children are complex, and I suggest that they do not lend themselves to accurate modelling.

Unreported discharges

Since the publication of the Black Report in July 1984,[8] there has been considerable speculation concerning the possibility that significant releases of radiation from Sellafield have gone unrecorded or undetected. This has been fuelled by the disclosure of previously unavailable information regarding a substantial release from the plant in the early 1950s.[17] As the *British Medical Journal* commented: 'This uncertainty means that it is impossible to exclude environmental radiation as a contributory cause of the cases of leukaemia observed in Seascale.'[18]

Metabolism

Ingestion

The metabolic models and parameters used by the NRPB in this stage of the calculation were essentially those recommended by ICRP.[19] Most of the metabolic parameters are based on experiments in animals, the results of which cannot reliably be extrapolated to humans.[20] With regard to the fractional uptake of actinides through the wall of the gastrointestinal tract, in animal experiments uptake varies by several orders of magnitude according to the species of experimental animal, the period of fasting before exposure, the rate of ingestion, and diet.[21–23] Unrealistically high concentrations of actinides are used simply to obtain measurable levels of uptake.[24]

These experiments usually employ radionuclides in their pure forms. It is known, however, that chemical complexing of actinides with organic materials in the environment may lead to increased absorption[24]. The only evidence concerning gut absorption of actinides in humans relates to ingestion of contaminated reindeer liver in Lapland and Finland, and of shellfish in Cumbria.[25,26] In the former study, no attempt was made to quantify large uncertainties in the estimates of Pu intake and in the calculations of whole-body content. It has been pointed out, furthermore, that neither reindeer liver nor shellfish are particularly typical components of the human diet, and thus in assessing values for gut transfer factors the data must be treated with caution.[27] Larsen reports that there is strong evidence that the absorption factor used in the Seascale risk assessment may have been 5–10 times too low.[21] In the recent amendment to its original calculation, the NRPB decided to double its estimate of gut absorption.[10] A recent reanalysis of the Finnish data by Leggett, however, suggests that this is still two times too low.[28] In short, the question of gastrointestinal absorption remains 'a hotly disputed topic'.[12]

Inhalation

Similar problems surround models of inhalation. There have been few long-term studies of particle retention in the human lung. Animal experiments are contradictory. For example, upon inhalation a fraction of the radioactive particles will be retained in the conducting airways and lymph nodes, and the remainder will be cleared from these sites to the gut by mechanical transport up the ciliary escalator. Studies on rats put the latter fraction at no more than 1 per cent, whereas those on dogs indicate that it could be as high as 15 per cent.[29] The ICRP assumes that 40 per cent of insoluble particles deposited in the pulmonary regions are cleared by mechanical transport with a half-time of one day.[19] Bailey, Fry and James, however, report finding no evidence for such a rapid clearance mechanism.[30] Newton, Taylor and Eakins have reported retention half-times for Pu and Am oxide in the lungs of a single worker 30 times higher than assumed by ICRP.[31] Far less material was cleared to the gut than would be predicted in the model. As a result, they concluded that ICRP's clearance model 'can grossly underestimate the long-term irradiation of systemic sites'.

To summarize my argument in this brief discussion of metabolism, the human biology of actinide metabolism is complex, and models remain crude and liable to substantial revision. I shall return to the question of child and fetal metabolism below.

Dosimetry

This brings me now to the question of alpha dose to radio-sensitive tissue. Only a small sphere of cells is irradiated around an alpha-emitter, and thus it may be important to pinpoint the location of leukaemia-sensitive cells in relation to actinide particles in the body.[32] The necessary information is simply not known; the NRPB assumed that leukaemogenic cells were situated in the red bone marrow (RBM) and distributed uniformly throughout its volume. This is a simplifying assumption. First, the leukaemias observed amongst Seascale children were predominantly leukaemias of the lymphatic variety. Lymph nodes and lymphatic tissues are widely spread throughout the body and are present within many tissues. Since insoluble radioactive particles are known to concentrate in the lymph nodes in the lung, it has been suggested that irradiation of these nodes might represent a leukaemic risk.[33] In experiments on animals, leukaemia has not been induced in the lungs of dogs or mice, but it has in rats.[34] Sir Edward Pochin of the NRPB has remarked in this regard: 'I don't know whether we are closer to the dog, the mouse, or the rat in terms of lymph node behaviour'.[34]

To return to the RBM, since the official inquiry into the Sellafield issue in 1984, the location of the sensitive cells within the marrow has come under close scrutiny.[33,35] There is evidence in animals that these cells—thought to be the haematopoietic stem cells—may be concentrated near the endosteal layer and are therefore more likely to be irradiated by bone surface-seeking radionuclides such as the actinides.[36] The situation is complex, since the lymphatic cells may be concentrated near the centre of the marrow; on the other hand, it is possible that their leukaemogenicity increases with proximity to the bone.[37] Thorne has estimated that considerations of microdosimetry contribute an uncertainty in RBM risk from the actinides of an order of magnitude.[12]

The thesis that the raised incidence of leukaemia amongst young children in Seascale could have been caused by exposure to actinides has recently been criticized by the NRPB on the grounds that long experience with isotopes of *radium* in humans has shown that the risk of radiation-induced cancer is principally a risk of bone cancer—osteosarcoma—and leukaemia is observed only at a far lower incidence.[10] Experience with Ra-224 is thought to be particularly relevant.[38] This isotope is assumed to deposit initially on the bone surfaces before it translocates to the bone volume. Because of its short half-life (3.65 days), it is assumed to decay at the bone surfaces *before* translocation. Ra-224 is thus taken to mimic the dosimetry of plutonium.[19]

This argument is based on observations of adult dial workers in the US exposed to Ra-226, and of German ankylosing spondylitis and tuberculosis sufferers—adults and children—administered with Ra-224; in each case the incidence of osteosarcoma exceeds that of leukaemia by an order of magnitude.[39–41] In the most pertinent case—that of children injected with Ra-224—no leukaemia has been observed whatsoever. On the other hand, of the 200 children given the radium therapy, 36 have developed bone cancer.[41] How, one might conclude, can plutonium be a leukaemic hazard?

This argument has two components:

(1) Bone cancer *is* observed in the children given Ra-224, and therefore, if environmental Pu is a significant risk to Seascale children, they should have developed bone cancer and *not* leukaemia.
(2) Leukaemia is *not* observed in the Ra-224 children, and therefore children exposed to Pu should also *not* develop leukaemia.

Consider the first of these arguments. Osteosarcoma is thought to originate in the osteogenic cells that lie in the endosteal layer on the inner bone surfaces, whereas leukaemia is assumed to originate within the RBM. Plutonium is taken to be a bone *surface*-seeker—that is, it is deposited on bone surfaces whence it irradiates the endosteum and bone marrow close to the endosteum. Radium, in contrast, is considered a bone *volume*-seeker—it distributes itself uniformly throughout bone material. As noted above, however, it is assumed by ICRP that Ra-224 mimics the dosimetry of Pu. Two pieces of evidence undermine this supposition. Firstly, it is possible that Ra does not decay at the bone surfaces but translocates rapidly to the bone volume. In experiments on rats a fairly uniform volume distribution is observed only one hour after injection.[42] In guinea-pigs it is reported that the volume deposit concentrates towards the bone surfaces.[43] It is not known whether human biology more closely resembles that of the rat or the guinea-pig.

Secondly, there is the question of the behaviour of Pu itself within the bone. The ICRP model makes the assumption that Pu in the skeleton is uniformly spread in an infinitely thin layer over endosteal bone surfaces; the fractions of the alpha-energy deposited in the endosteum and in the RBM are taken to be equal.[19] The biology of Pu in bone, however, is far more complex. In 1979, Priest and Hunt proposed a revision to the ICRP model that would allow for the burial and recycling of Pu within the skeleton.[44] Pu deposited in the skeleton is continuously buried by the apposition of new bone and continuously removed into the marrow itself by bone resorption: it is 'scavenged' by the macrophages. This cycle is endlessly repeated with a period in adults of about a decade.

According to this revised model, the Pu-239 dose to the RBM is almost unchanged from that calculated by the ICRP model. On the other hand, the dose to the endosteum is reduced by a factor of 30. Thus the authors concluded that, following intakes of Pu-239, leukaemia could in fact be *more* common than osteosarcoma.[44] In 1984, Priest and Birchall further revised the model and suggested that the gap between endosteal and marrow dose might not be so large.[45] These kinds of empirical and theoretical considerations demonstrate, however, that it would be premature to assume either that Pu behaves like Ra, or that skeletal Pu should give rise to osteosarcoma in preference to leukaemia. There is clearly scope for models to be revised in the light of empirical experience.

Consider now the second component of the original argument, namely that no leukaemias were observed amongst the children administered with Ra-224, and therefore Pu should not cause leukaemia. Surely, some part of the marrow must have been irradiated, so why were no leukaemia cases observed? This is all the more surprising since, amongst the corresponding adults given Ra-224, two cases of leukaemia were observed instead of the one expected;[40] children are generally assumed to be *more* sensitive to radiation-induced leukaemia than adults. Surveying this apparent enigma, in 1976 Thorne and Vennart noted that the skeletal doses to the children were massive—roughly an order of magnitude greater than those received by the adults[46] (Table 1). They suggested that at such high doses a cell-killing effect might account for the absence of leukaemia in the children. In other words, they postulated a bell-shaped dose–response for radiation leukaemogenesis.

For this hypothesis to be seriously entertained, it would be necessary to know whether the induction of leukaemia is suppressed by amounts of alpha-emitter that still give rise to osteosarcoma. Recent work at the Medical Research Council Radiobiology Unit set out specifically to test this idea. Humphreys *et al.* gave injections of Ra-224 to CBA mice, a strain with an extremely low natural incidence of myeloid leukaemia (there has been no recorded case in 800 untreated animals so far observed for their lifespan).[47] First, these authors found that Ra-224 induces myeloid leukaemia in CBA mice (Table 2). Secondly, the frequency of leukaemia induction falls off as the frequency of osteosarcoma begins to rise. Using simple mathematical models to quantify this relationship, the authors represented their results graphically (Figure 2). As can be seen, there is a well-defined peak in leukaemia induction just before the osteosarcoma incidence begins to rise rapidly.

These results, the authors concluded, are thus consistent with the hypothesis that amounts of bone-seeking alpha-emitter sufficient to induce osteosarcoma may also be sufficient to sterilize those cells in the marrow that have leukaemogenic potential.[47] It remains to be established whether the distribution and sensitivity of leukaemogenic cells in humans are the same as in CBA mice.

Table 1 German TB and ankylosing spondylitis patients injected with Ra-224: average exposure parameters[40]

	Sex	No. of patients	Age at first injection	Follow-up (yrs)	Injection Ra224 (μCi/Kg)	Skeletal dose (rads)
Adults	M	455	39	17	13.3	186
	F	157	37	17	20.2	282
Children	M	103	11.6	20	30.7	1182
	F	101	11.7	19	27.3	1038

Table 2 Incidence of myeloid leukaemia and osteosarcoma in male CBA mice. (Single and multiple injection experiments)[47]

224-Ra (kBq)	0	2	4	8	16	32	64
Single injection experiment							
Number of mice	41	41	41	41	41	41	41
Number with myeloid leukaemia	0	0	0	2	3	1	1
Number with osteosarcoma	0	0	0	1	1	1	3
Multiple injection experiment							
Number of mice	36	40	41	42	39	39	41
Number with myeloid leukaemia	0	0	1	2	1	0	0
Number with osteosarcoma	0	1	1	0	1	2	4

To summarize this brief discussion of dosimetry, the considerations relating to Pu in bone are complex and models are open to potentially significant revision in the light of empirical experience.

Child and fetal models

Before I move on to the question of dose response, I should point out that, with very minor exceptions, all the parameters in the metabolic and dosimetric

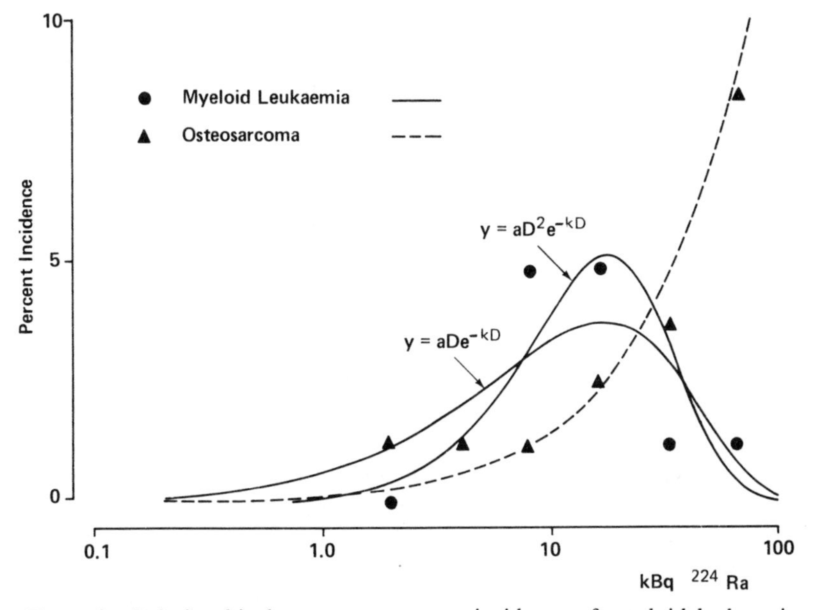

Figure 2 Relationship between percentage incidence of myeloid leukaemia and osteosarcoma in male CBA mice and amount of injected Ra-224. (Combined single and multiple injection experiments.)[47]

models employed in the risk assessment were those recommended by the ICRP for *adult* workers. The only significant alteration to the adult *models* was the introduction of a factor to account for the reciprocal relationship between tissue dose and the increasing body mass of the growing child; thus the dose—and the risk arising—from a long-lived alpha-emitter lodged in the skeleton was taken to *decrease* with the growth in child body mass.[48] As I have already argued, in an assessment of risk from actinides, tissue mass is relatively unimportant, and at the cellular level alpha-particle dose averaged throughout the tissue may be biologically almost meaningless. The NRPB's models of actinide metabolism and dosimetry in the fetus were similarly concerned mainly with the growing fetal mass: thus the fractional distribution of Pu in the organs of the fetus was taken to be the same as that in the mother.[49]

It is a general shortcoming of the risk assessment that uncertainties such as the effect of possible age-specific metabolism were given only a *post hoc* and piecemeal treatment at the end of the calculation; there was no attempt to propagate uncertainties through the assessment in a formal and systematic manner. Thus the final risk figure was given without an error bar and without consideration of uncertainty owing to separate errors in combination. This feature of the assessment is itself an indicator of the extent of uncertainty: the sorts of considerations I have outlined do not lend themselves to formal statistical techniques of error analysis. Under such conditions, the best that can be achieved are order-of-magnitude estimates based on the range of possible answers.[50]

These conditions obtain in particular with respect to the modelling of doses to young children and the fetus. The sensitivity analysis in the Seascale risk assessment relied at best on a tiny handful of animal experiments; in the case of the fetus extreme care is needed in the extrapolation from animals to humans.[49] The uncertainties are so broad and uncharted that it is difficult to estimate their magnitude or the specific sites at which they are most likely to be significant.

Dose–response

When discussing the extent of scientific controversy over this question, it is hard not to lapse into hyperbole. The literature is littered with examples of sharp disagreement. In 1980, roughly a factor of 50 separated Harold Rossi and Edward Radford on the BEIR-III Committee;[51,52] their disagreement persists today. In 1979, the late G. W. Dolphin of the NRPB noted that the ICRP system of radiological protection relies on the *precise* estimation of two parameters: the level of risk acceptable to an exposed population, and the cancer risk per unit dose. He wrote that '[this] leads to an intellectually satisfying system of radiological protection, provided disbelief in the values of the 2 parameters is suspended'.[53]

Of course, there is controversy over the extent to which young children and the fetus are sensitive to radiation-induced leukaemia, but uncertainty over dose–response for *alpha*-radiation is particularly acute.[54] The re-estimation of the T65 neutron doses at Hiroshima and Nagasaki casts a further shadow of doubt over the largest and most carefully researched body of data available.[55] A number of laboratory studies indicate possible supra-linearity at low doses, implying that the linear interpolation for high LET may involve an underestimation of the risk;[56–58] it is now well established that the relative biological effectiveness of alpha-radiation depends also on the biology of the target tissue.[58] It has been remarked that the accumulation of evidence in this area of cancer biology at present serves only to illuminate our ignorance.[57]

The problem of dose–response is neatly summed up in a recent journal article:

> Since there is judgement as well as technical evaluation from the data, it is inevitable that there will be dissent from the values put forward. There is no evidence to disprove the ICRP risk factors, however, neither is there any evidence to verify them.[59]

SOCIAL PROBLEMS OF RISK ASSESSMENT

Now I shall offer some approximate figures for the uncertainties I have described. The uncertainty in exposure is significant, but less than an order of magnitude; uncertainties in metabolism and dosimetry of actinides I consider to be in each case in excess of an order of magnitude; as for dose–response, the years of debate have whittled away at the uncertainty, which I put at about a factor of 5. Thus the final uncertainty in the Seascale risk assessment is likely to be between three or four orders of magnitude.

I have used the example of Seascale somewhat rhetorically to present an argument that I consider has general application to the problems of environmental radiation risks. My conclusion is that alpha-emitters discharged from Sellafield manifestly *could* have been the cause of the observed leukaemia incidence. This is far from saying that they *were* the cause: this we can never know for certain. The problem, however, is not only the pervasive scientific uncertainty in modelling population doses, but the problem of adapting institutions and changing public consciousness so that regulatory mechanisms can operate effectively under the conditions of such uncertainty. How can we develop technologies, the risks of which we do not understand? This is a general problem common to many large-scale chemical and industrial risks, and not only to radiation.

In these circumstances, the modelling of radiation risks can provide only a loose and flexible framework within which *political* debate between parties in dispute over nuclear policies can take place. The most important question is then no longer 'what is the precise risk?', but 'who benefits from the doubt?' This is a problem of accountability and control.

There are many influences acting against the formal recognition and incorporation of uncertainty considerations into the regulatory system. In the case of radiological protection, some of these can be enumerated as follows.

Regulatory-scientific

Of all the hazardous substances in the environment, low-level radiation has been the most extensively researched. Scientists and regulators in the chemical industry look towards the ICRP system of radiological protection as a paragon of regulatory virtue. In chemical toxicology, a 'rad-equivalent' chemical dose is often concocted in order to evoke the authority of radiological protection standards. One can detect the note of despair in the voice of the Director of the British Industrial Biological Research Association when confronted by the uncertainty in dose–response for radiation carcinogenesis:

> if you, in the radiation industry, if I may put it like that, cannot define more accurately the lower limits of the dose response curve, so that you are forced into using the straight-line theoretical extrapolation, what hope is there for the chemical industry?[60]

Thus there are pressures from within the scientific-regulatory system to stress the (relatively) advanced scientific status of radiation risk assessments.

Pressures against the recognition of uncertainty also arise from and are built into the ICRP system itself. For example, precision is the *sine qua non* of optimization and the intercomparison of risks. As ICRP 26 explains, the more 'cautious and conservative' the assessment of risk (i.e. the more explicit the recognition of uncertainty),

> the more important it becomes to recognise that it may lead to an overestimate of the radiation risk, which in turn could result in the choice of alternatives that are more hazardous than practices involving radiation.[61]

The former Director of the MRC Radiobiology Unit, Dr R. H. Mole, has noted that

> Throughout the last two decades there seems to have been continuing pressure to complete the construction of a completely comprehensive and internally consistent system of radiation standards in the belief that this was practicable and ought to be made available.[32]

Recognizing the uncertainty, however, Mole expressed reservations: 'My personal concern is that tidyness for regulatory reasons may serve to conceal ignorance and confirm complacency'.[32]

Since its inception, ICRP has striven towards the mathematization of its radiological protection system;[62-64] the desire for regulatory 'tidyness' is an obvious feature of radiological protection in the UK and abroad.

Regulatory-institutional

It is undoubtedly very difficult for scientific institutions with regulatory responsibilities to operate in full recognition of the uncertainties that underlie their tasks. Thus we can expect pressure towards a technocratic faith in the efficacy of a radiological protection system to arise from *within* scientific institutions themselves. Wynne has noted in this regard:

> The intrinsic commitments of [nuclear] technologies are not only material; they are also organisational, intellectual, and psychological. The people involved need not only to have largely taken-for-granted interpretive frameworks to act as the foundation for working rules or practices, but also to have total faith, unsullied by doubts and uncertainties, in the success and worth of the technology. ... Organisational morale becomes a matter of crucial importance to concentration and coherence, and much management effort goes in symbolic action, in image-building and projection to sustain the faith. Objections and criticisms tend to be treated as illegitimate and irrational, rather than as legitimate equals in debate: uncertainties and accidents are downplayed as under control.[4]

Regulatory-scientific institutions must also have an external, 'public' face. An aura of certainty is almost a prerequisite for their authority to be upheld;[5] it is all too often assumed that 'experts' are simply repositories for the 'truth'. Moreover, there exist *inter*-institutional pressures to play down uncertainty. To put it somewhat crudely, 'objective' standards and precise numbers are often demanded by industry (so it can plan) and by the public (so it can take industry to court). Neither regulatory agencies nor other actors in the system can easily operate with an open recognition of the uncertainties involved.[6]

SUMMARY AND CONCLUSIONS

Public confidence in radiological protection is low. In the first part of this chapter I have shown that confidence cannot be re-established through an evocation of precise certainty in radiation-risk assessment: the scientific basis for such precision and certainty does not exist. Yet public, scientific and regulatory consciousness is not attuned to the extent of the scientific ignorance on these matters. In the second part of the chapter I have described some of the pressures acting to maintain such technocratic illusions.

In the first place, it is important to recognize that, if the first part of the analysis is correct, the distinction between 'expert' and 'layperson' no longer holds good. If they are to gain credence, scientists have to *persuade* and to *argue*, and not merely to inform. An adversarial position cannot be neutral, and controversy is likely to polarize around essentially political positions.[65] In these circumstances, scientific arguments are only one element within general debates, and 'experts' should not expect to enjoy any privileged or arbitrating role.[66]

Secondly, any system of radiological protection that recognizes uncertainty must be able to respond to the inevitable mistakes and disputes which will arise as a result of practical experience with nuclear technology. Once again, the resulting problems will be of an essentially political nature, and in these circumstances the role of the scientifically trained can only be to advise, and not to dictate. Under these conditions, decision-making will have to be open to participation of the public and workers whose interests are at stake; political problems cannot be resolved by scientific fiat.

The second part of my analysis suggests that such a democratization of decision-making will face many problems—psychological, social and institutional. Finally, there are likely also to be intense *political* pressures at work to resist the realization of such a scenario.[67]

REFERENCES

1. A. M. Weinberg, Science and trans-science, *Minerva*, **X**, 209–22 (1972).
2. R. Johnston, 'The characteristics of risk assessment research', in *Society, Technology and Risk Assessment*, (ed. J. Conrad) Academic Press, New York, 1980, pp. 105–22.
3. B. Wynne, 'Technology, risk and participation: on the social treatment of uncertainty', *Society, Technology and Risk Assessment*, (ed. J. Conrad), Academic Press, New York, 1980, pp. 173–208.
4. B. Wynne, Redefining the issues of risk and public acceptance: the social viability of technology, *Futures*, **15**, 13–32 (1983).
5. G. Majone, Process and outcome in regulatory decision-making, *Am. Behav. Sci.*, **22**, 561–83 (1979).
6. J. Linnerooth, The political processing of uncertainty, *Acta Psychol.*, **56**, 219–31 (1984).
7. D. Crouch, Science and trans-science in radiation risk assessment: child cancer around the nuclear fuel reprocessing plant at Sellafield, U.K., Sci. *Total Environ.*, **53**, 201–16 (1986).
8. *Investigation of the Possible Increased Incidence of Cancer in West Cumbria*, Report of the Independent Advisory Group, HMSO, London, 1984.
9. National Radiological Protection Board, The Risks of Leukaemia and Other Cancers in Seascale from Radiation Exposure, *NRPB-R171*, Didcot, 1984.
10. Addendum to R171, NRPB, Didcot, 1986.
11. H. J. Dunster, Discharges from Sellafield, *Lancet*, **2**, 873 (1984).
12. M. C. Thorne, *Critique of NRPB Estimate of Tissue-specific Radiation Exposure of the Seascale Population*. Paper presented to joint meeting of Royal Statistical Society and Society for Social Medicine: Public Controversies and Scientific Evidence, Royal Society of Medicine, London, 25 June 1985.

13. G. Hunt, Timescales for dilution and dispersion of transuranics in the Irish Sea near Sellafield, *Sci. Total Environ.*, **46**, 261–78 (1986).
14. J. D. Eakins and A. E. Lally, The transfer to land of actinide-bearing sediments from the Irish Sea by spray, *Sci. Total Environ.*, **35**, 23–32 (1984).
15. E. I. Hamilton and K. R. Hunt, The recent sedimentation history of the Esk estuary, Cumbria, UK, *Sci. Total Environ.*, **35**, 325–86 (1984).
16. J. R. Simmonds, J. Stather and F. Fry, *Levels of Exposure and Radiation Risks at Seascale.* Paper presented to joint meeting of Royal Statistical Society and Society for Social Medicine: Public Controversies and Scientific Evidence, Royal Society of Medicine, London, 25 June 1985.
17. D. Jakeman, Notes on the Level of Radioactive Contamination in Sellafield Areas Arising from the Discharges in the Early 1950s, *AEEW R 2104*, London, HMSO, 1984.
18. Anon., New estimates of radioactive discharges from Sellafield, *Brit. Med. J.*, **293**, 340 (1986).
19. International Commission on Radiological Protection, Limits on intakes of radionuclides by workers, *ICRP Publication No. 30*, Parts 1–3, Ann. ICRP, 2(3/4), 4(3/4), 6(2/3), 1979–1981.
20. *The Effects on Populations of Exposure to Low Levels of Ionizing Radiation.* Report of the Committee on the Biological Effects of Ionizing Radiation, National Research Council, Washington, DC, 1980, p. 28.
21. R. P. Larsen, D. M. Nelson, M. H. Bhattacharyya and R. D. Oldham, Plutonium—its behaviour in natural water systems and assimilation by man, *Hlth Phys.*, **44** Suppl. 1, 485–92 (1983).
22. M. F. Sullivan, B. M. Miller and L. S. Gorham, Nutritional influences on plutonium absorption from the gastrointestinal tract of the rat, *Radiat. Res.*, **96**, 580–81 (1983).
23. J. D. Harrison, The gastrointestinal absorption of plutonium, americium and curium, *Radiat. Prot. Dosimetry*, **5**, 19–35 (1983).
24. J. R. Cooper and J. D. Harrison, The speciation of plutonium in foodstuffs and its influence on gut uptake, *Sci. Total Environ.*, **35**, 217–25 (1984).
25. H. Mussalo-Rauhamaa, T. Jaakola, J. K. Miettinen and K. Laiho, Plutonium in Finnish Lapps: an estimate of the gastrointestinal absorption of plutonium by man based on a comparison of the plutonium content of Lapps and Southern Finns, *Hlth. Phys.*, **46**, 549–59 (1984).
26. G. J. Hunt, D. R. P. Leonard and M. B. Lovett, Transfer of environmental plutonium and americium across the human gut, *Lancet*, **1**, 439–40 (1986).
27. D. M. Taylor, Gut transfer of environmental plutonium and americium, *Lancet*, **1**, 611 (1986).
28. R. W. Leggett, An upper-bound estimate of the gastrointestinal absorption fraction for Pu in adult humans, *Hlth. Phys.*, **49**, 1299–1301, (1985).
29. M. R. Bailey, F. A. Fry and A. C. James, Long-term retention of particles in the human respiratory tract, *J. Aerosol. Sci.*, **16**, 295–305 (1985).
30. M. R. Bailey, F. A. Fry and A. C. James, The long-term clearance kinetics of insoluble particles from the human lung, *Ann. Occup. Hyg.*, **26**, 273–90 (1982).
31. D. Newton, B. T. Taylor and J. D. Eakins, Differentials clearance of plutonium and americium oxides from the human lung, *Hlth Phys.*, **44**, Suppl. 1, 431–9 (1983).
32. R. H. Mole, The biological basis of plutonium safety standards, *J. Br. Nucl. Energy Soc.*, **15**, 203–213 (1976).
33. J. Cutler, Sellafield, Seascale, and stem cells, *Lancet*, **2**, 1161 (1984).

34. E. E. Pochin, *Sizewell Inquiry Transcripts*, day 151, 101, (1983).
35. J. Cutler, Windscale: close to the bone, *New Statesman*, 1 February, 13 (1985).
36. B. I. Lord and E. G. Wright, Spacial organization of CFU-S proliferation regulators in the mouse femur, *Leuk. Res.*, **8**, 1073–83 (1984).
37. Patterson Laboratories, Christie Hospital, Manchester, July 1985, personal communication.
38. R. H. Mole, *Biological and Radio-biological Aspects of Leukaemogenesis by Ionizing Radiation, with Special Reference to Children at Low Levels of Exposure*. Paper presented to joint meeting of the Royal Statistical Society and Society for Social Medicine: Public Controversies and Scientific Evidence, Royal Society of Medicine, London, 25 June 1985.
39. F. W. Spiers, H. F. Lucas, J. Rundo and G. A. Anast, Leukaemia incidence in the US dial workers, *Hlth. Phys.*, **44S1**, 65–72 (1983).
40. H. Spiess, A. Gerspach and C. W. Mays, Soft-tissue effects following 224-Ra injections into humans, *Hlth. Phys.*, **35**, 61–81 (1978).
41. C. W. Mays, H. Spiess and A. Gerspach, Skeletal effects following 224-Ra injections into humans, *Hlth. Phys.*, **35**, 83–90 (1978).
42. N. D. Priest, G. Howells, D. Green and J. W. Haines, Autoradiographic studies of the distribution of Ra-226 in rat bone—their implications for human dosimetry and toxicity, *Hum. Toxicol.*, **2**, 479–96 (1983).
43. N. D. Priest, Autoradiographic studies of the distribution of Radium-226 in rat and Guinea pig bones, *Calcif. Tiss. Int. Suppl.*, **35**, A31 (1983).
44. N. D. Priest and B. W. Hunt, The calculation of annual limits of intake for Plutonium-239 in man using a bone model which allows for Plutonium burial and recycling, *Phys. Med. Biol.*, **24**, 525–46 (1979).
45. N. D. Priest and A. Birchall, The calculation of bone doses from alpha-emitting bone surface seeking radionuclides for radiological protection purposes. *Radiat. Environ. Biophys.*, **23**, 149–53 (1984).
46. M. C. Thorne and J. Vennart, The toxicity of 90-Sr, 226-Ra and 239-Pu, *Nature*, **263**, 55–8 (1976).
47. E. R. Humphreys, J. F. Loutit, I. R. Major and V. A. Stones, The induction by 224-Ra of myeloid leukaemia and osteosarcoma in male CBA mice, *Int. J. Radiat. Biol.*, **47**, 239–47 (1985).
48. N. Adams, Dependence on age at intake of committed dose equivalents from radionuclides, *Phys. Med. Biol.*, **26**, 1019–34 (1981).
49. N. Adams and J. W. Stather, Irradiation of the foetus from maternal intakes of plutonium, *Radiol. Prot. Bull.*, **58**, 31–6, (1984).
50. M. G. Morgan, 'Uncertainty and quantitative assessment in risk management', in *Assessment and Management of Chemical Risks*, (ed. J. V. Rodricks and R. G. Tardiff), ACS Symp. 239, American Chemical Society, Washington, DC, 1984, pp. 113–29.
51. H. H. Rossi, Comments on the somatic effects section of the BEIR-III report, *Radiat. Res.*, **84**, 395–406 (1980).
52. E. P. Radford, Human health effects of low doses of ionizing radiation, *Radiat. Res.*, **84**, 369–94 (1980).
53. G. W. Dolphin, Radiation carcinogenesis and radiological protection, *Environ. Res.*, **18**, 140–46 (1979).
54. M. W. Charles and P. J. Lindop, Risk assessment without the bombs, *J. Soc. Radiol. Prot.*, **1**, 15–19 (1981).
55. W. E. Loewe and E. Mendelsohn, *Revised Estimates of Dose at Hiroshima and*

Nagasaki and Possible Consequences for Radiation Induced Leukaemia, Lawrence Livermore Laboratory Report LLNL D-80 14, 1980.
56. J. M. Brown, Linearity vs non-linearity of dose response for radiation carcinogenesis, *Hlth Phys.*, **31**, 231–45 (1976).
57. J. E. Coggle, Dose effect relationships for radiation induced cancer: relevance of animal evidence, *J. Soc. Radiol. Prot.*, **2**, 15–21 (1982).
58. R. J. M. Fry, Experimental radiation carcinogenesis: what have we learned? *Radiat. Res.*, **87**, 224–39 (1981).
59. J. A. Reissland, Radiation epidemiology, *J. Soc. Radiol. Prot.*, **3**, 29–33 (1983).
60. D. M. Conning, *Arch. Toxicol. Suppl.*, **3**, 52 (1980).
61. International Commission on Radiological Protection, Recommendations of the International Commission on Radiological Protection, *ICRP Publication No. 26*, Ann. ICRP, 1(3) 1977.
62. R. Bertell, *No Immediate Danger: Prognosis for a Radioactive Earth*, London, Women's Press, 1985, pp. 172–4.
63. R. Bertell, *The Human Health Consequences of Exposure to Ionizing Radiation*, Proof of Evidence SSBA/P/8, Sizewell Inquiry, 1984.
64. P. A. Green, *The International Commission on Radiological Protection*, London, Greenpeace, 1985.
65. M. Thompson, Among the energy tribes: a cultural framework for the analysis and design of energy policy, *Policy Sci.*, **17**, 321–39 (1984).
66. D. Nelkin, The political impact of technical expertise, *Soc. Stud. Sci.*, **5**, 35–54 (1975).
67. D. Crouch, Hazards of estimating hazards: radiation risk assessment in the U.K., unpublished, Science Policy Research Unit, Brighton, 1986.

Radiation and Health
Edited by R. Russell Jones and R. Southwood

Published 1987 by John Wiley & Sons Ltd.

5

Assessing Risks of Childhood Leukaemia in Seascale

J. W. STATHER, J. DIONIAN, J. BROWN, T. P. FELL and C. R. MUIRHEAD
National Radiological Protection Board, Chilton, Didcot, Oxon, UK.

ABSTRACT

In the village of Seascale (population about 3000), which is situated approximately 3 km to the south of the Sellafield nuclear fuel reprocessing plant in west Cumbria, four fatal leukaemias have been observed in children under 20 years of age between 1950 and 1980. Based on UK statistics only 0.5 leukaemias would have been expected. Because of public concern that these leukaemias could have resulted from discharges of radioactive materials from the Sellafield site, radiation doses and risks of radiation-induced leukaemia have been calculated for children born and living in Seascale over the period of operation of the plant.

A study by the Board has shown that for the Seascale study population of 1225 children and young persons, followed to age 20 years, or 1980 for those born after 1960, 0.016 radiation-induced leukaemias would be expected from the Sellafield discharges. This corresponds to an average risk to children in the Seascale population of about 1 in 75 000. For the four fatal leukaemias observed in the study population to be attributed to the operations at Sellafield, the average risk would have to be increased by a factor of about 250, to 1 in 300. Although there is some uncertainty about the releases from the plant and concentrations of radionuclides in foodstuffs in the Sellafield area, particularly for the early years of its operation, the possibility that the doses calculated and the risk coefficients used for radiation-induced leukaemia could be so substantially wrong is very unlikely.

The number of radiation-induced leukaemias in the study population, from all radiation sources, is calculated to be 0.1, which corresponds to a risk of about 1 in 12 250 for the average child. About two-thirds of the risk is calculated to result from natural radiation and 16 per cent from the Sellafield discharges with nuclear weapons fallout and medical exposure each contributing about 9 per cent.

THE SEASCALE STUDY POPULATION

Seascale is a small coastal village with a population of about 3000 situated approximately 3 km to the south of the Sellafield (Windscale) nuclear fuel reprocessing plant operated by British Nuclear Fuels plc (BNFL). It has been observed that the incidence of childhood leukaemia in the village is approximately ten times that expected on the basis of UK national statistics,[1] and inevitably the possibility has been raised that this was the result of discharges into the local environment of radioactive materials from the Sellafield plant.

A Government Enquiry was set up in November 1983 to look into the possible risks to children living in the area and the NRPB was asked to review all possible routes of exposure of children and young persons in Seascale to discharges from the plant, and to calculate radiation doses and risks. The Board produced an initial report in July 1984[2] which was based on information on discharges that was readily available at that time, although it was recognized that some information on releases from the plant, particularly in the early years of its operation, were very limited. In the following 18 months, BNFL undertook a comprehensive review of all plant operating data and this resulted in a number of modifications to reported discharges. A specific aspect of some of this new information related to the release of particles of irradiated uranium fuel from the Windscale piles during the 1950s, and arose from a query by Dr Jakeman (AEE Winfrith) who worked in Research and Development at Windscale at that time. This new information was included in a second report which was issued in July 1986.[3]

This chapter outlines the methods adopted for the analysis, together with the main results and conclusions.

ASSESSMENT PROCEDURE

The aim of the study was to estimate radiation doses and risk of leukaemia to children and young persons born and living in Seascale over the period of operation of the Sellafield plant up to 1980. Figure 1 shows the assessment procedure used. The study was mainly concerned with calculating risks associated with the Sellafield discharges, but all the main sources of radiation exposure of the Seascale population were considered in the analysis. These are:

(a) Releases of radioactive materials both to sea and to atmosphere from the routine operations at the Sellafield plant operated by BNFL and as a consequence of accidents and incidents; the Windscale fire in 1957, was treated as a separate source.
(b) Fallout from the atmospheric testing of nuclear weapons.

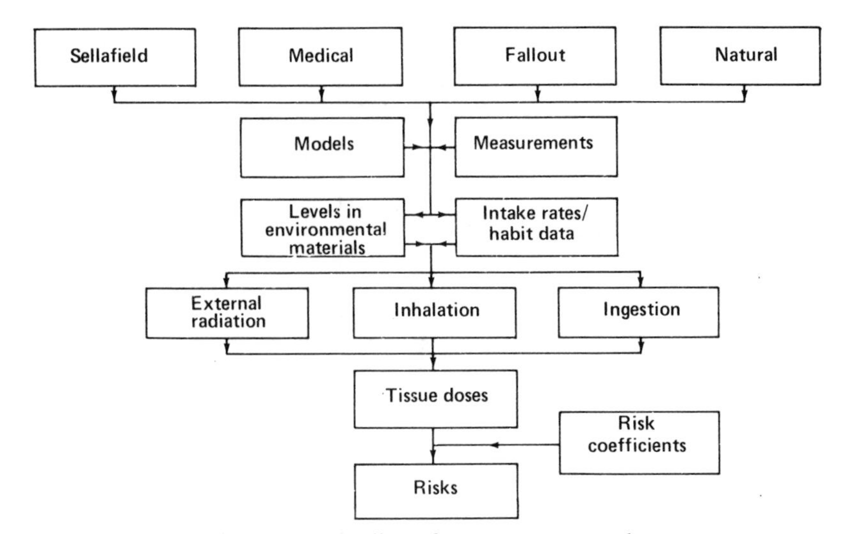

Figure 1 Outline of assessment procedure.

(c) Natural radiation.
(d) Diagnostic medical practices.

As far as possible the assessment was based on measured values of the appropriate parameters (external dose rates, concentrations of radioactive materials in foodstuffs and air etc.). Where measurements were not available the discharge data were used to obtain estimates of the appropriate parameters. Intake rates and habit data were based on published reports where possible. They were used to calculate intakes of radionuclides by inhalation and ingestion pathways as well as external radiation exposure, and hence doses to the red bone marrow and risks of radiation-induced leukaemia.

DISCHARGES FROM THE SELLAFIELD SITE

As a result of the operations at the Sellafield plant, radioactive materials are discharged into the atmosphere and into the Irish Sea. The data on routine and accidental releases of radioactive material to the environment from the Sellafield site were obtained from BNFL and from the UK Atomic Energy Authority (AEA) and have been published.[3] As a consequence of these releases the Seascale population may be exposed to radiation by both marine and terrestrial pathways. The main pathways that have been identified and for which doses to the population were calculated are:

Discharges to atmosphere

External radiation from radionuclides in the air and deposited on the ground.
Inhalation of radionuclides in the atmosphere.

Table 1 Intake and occupancy data for children and young persons with average habits in the Seascale population

Age group (years)	Mean ingestion rate (kg y^{-1})								Beach occupancy (h y^{-1})	Breathing rate (m^3 y^{-1})[d]
	Fish[a]	Crustaceans[a]	Molluscs[a]	Milk[b]	Meat[c] + offal	Vegetables[c]	Sand	Soil		
0–2	2	0	0	170	3.2	35	0.001	0.037	25	1.4 10^3
3–7	6	0.2	0.2	150	15.1	51	0.002	0.037	50	4.8 10^3
8–14	7	0.3	0.3	150	18.3	60	0.002	0.018	100	5.5 10^3
15+	10	0.5	0.5	150	27	90	0.002	0.018	100	8.4 10^3

[a] 100 per cent assumed to have been caught locally.
[b] 40 per cent assumed to have been produced locally.
[c] 25 per cent assumed to have been produced locally.
[d] Ref. 26.

Ingestion of locally produced food (milk, vegetables and meat are produced locally; it was assumed that 40 per cent of milk and 25 per cent of the other foods were of local origin).
Inadvertent ingestion of soil, which occurs when children play outside and put their fingers into their mouths.

Discharges into the Irish Sea

External radiation from radionuclides deposited on the beach.
Inhalation of sea-spray and resuspended sand.
Ingestion of locally caught sea-food (fish, crustaceans and molluscs are an important part of the diet in coastal communities; it was pessimistically assumed that all sea-food consumed by the study population was caught locally).
Inadvertent ingestion of sand.

The mean rates for ingestion of terrestrial and marine foods, inadvertent ingestion of soil and sand, and inhalation are given in Table 1.

CONCENTRATIONS OF RADIOACTIVE MATERIALS IN ENVIRONMENTAL MATERIALS AS A RESULT OF DISCHARGES FROM SELLAFIELD

For the analysis, concentrations of radionuclides in the appropriate environmental materials were needed for the period of operation of the Sellafield plant. Very many measurements have been made over the years by BNFL, by various government departments and more recently by NRPB, of levels of radionuclides in environmental materials. Where measurements had been made of the appropriate parameters these were used in the analysis; where the data were not available concentrations were determined from the discharges using appropriate models.[2,3]

For example, measurements were generally available of levels in milk of the principal radionuclides discharged from the Sellafield plant. For the years when measurements were not available, the milk concentrations were calculated using a scaling factor relating measured concentrations in other years to the atmospheric discharges in those years. All measurements were corrected for any contribution from weapons fallout. Concentrations of radionuclides in other terrestrial foodstuffs were calculated using food-chain models developed at NRPB.[4,5] These models were used to obtain ratios of the concentrations of each radionuclide in the various foods considered compared with milk. These ratios were then used to estimate concentrations in other foods from the milk concentrations.

Measurements of radionuclides in marine foods were used where possible.

However, measurements in earlier years and for some foods and radionuclides are limited. Where measured concentrations were not available, these were estimated from the discharges. The relationship between concentrations in marine foods and discharges were derived by MAFF from measurements in recent years and are described elsewhere.[6]

Concentrations of radionuclides in the appropriate environmental materials between 1950 and 1982 have been published.[3]

HABIT DATA

To calculate intakes of radionuclides and external exposure, information was required on the habits of the Seascale population (breathing rates, how much fish is consumed, how much time people spend on the beach, etc.). There is a substantial amount of information in the literature for the UK population as a whole[7] but these data are not necessarily appropriate for a local coastal community. Local habit surveys have also been carried out around Seascale by Government Departments, but these concentrate on identifying the higher than average consumers for critical group calculations.[6,8] Again these data are not appropriate for the average Seascale resident. It was, therefore, necessary to use published data as guidelines to establish parameters for the study population. The study was concerned with children and young persons in Seascale and habit data were required for the four representative age groups considered in the calculations. These were 0 to 2-year-olds, 3 to 7-year-olds, 8 to 14-year-olds and adults. The habit data used in the analysis are shown in Table 1.

OTHER SOURCES OF RADIATION EXPOSURE OF THE POPULATION

Natural radiation

The principal sources of natural radiation are cosmic rays and terrestrial radionuclides, the most important of which are ^{40}K and the radionuclides in the decay series of ^{238}U and ^{232}Th. All materials in the earth's crust contain these radionuclides and as a consequence persons are continuously exposed to gamma-radiation from their radioactive decay. Persons are also exposed as a result of intakes of these radionuclides in air and diet. In calculating radiation doses from natural radiation the most up-to-date UK data were used, including that recently published by the NRPB.[2,9] When data were not available use was made of information published by UNSCEAR.[10,11]

Fallout from weapons testing

Measurements of fallout radionuclides in air, water and food, principally milk, have been made in the UK since the 1950s. The data used in the analysis were taken from published reports of the Radiobiological Laboratory of the Agricultural Research Council, The Ministry of Housing and Local Government, the United Kingdom Atomic Energy Authority and the NRPB. Where appropriate, national data were scaled according to the rainfall for the coastal area of west Cumbria. Doses were calculated for external radiation and for intakes of radionuclides by inhalation and ingestion.

Diagnostic medical exposure

Radiation and radionuclides are used in medicine for a variety of diagnostic procedures. Irradiation of the body is, in general, heterogeneous. Only data from very limited surveys are available,[12,13] and these surveys have shown that doses are quite variable. There are, therefore, likely to be substantial local differences in the radiation doses received. For this analysis there was no alternative but to apply these limited national data to the population of west Cumbria.

CALCULATION OF TISSUE DOSES

Internally incorporated radionuclides

A suite of computer programs, PEDAL (Programs Estimating Doses and Annual Limits), written to incorporate the dosimetric models and metabolic data recommended by ICRP[14] was used to calculate absorbed radiation doses in the red bone marrow from low LET and high LET radiation after intakes of radionuclides by inhalation and ingestion. Body and organ growth was allowed for.[15]

It was not practicable to attempt to calculate doses from all the radionuclides to which the Seascale population might be exposed. Radiation doses were therefore calculated for the 26 radionuclides estimated to give more than about 95 per cent of the dose to the red bone marrow from incorporated radionuclides.

The ICRP dosimetric models for the lung and gut[14] were assumed to apply to all ages. The gut transfer factors (f1) used for ingested radionuclides were those appropriate for transportable forms of the radionuclides. The only departure from the recommendations in ICRP Publication 30 was that the f1 used for plutonium, americium and thorium isotopes was 1×10^{-3}. This value has recently been recommended by an ICRP Task Group[16] as a cautious value to use for these actinides incorporated in foodstuffs. For inhalation, all radionuclides were treated as inhalation Class W unless they would normally be treated as Class D (e.g. Cs, I, Sr). This is a conservative approach for calculating doses to the red bone marrow. This assumption could potentially underestimate radiation doses to the lymph nodes draining the lung if a substantial fraction of the plutonium and americium inhaled behaved as lung Class Y (i.e. as insoluble material). However, the available experimental evidence suggests that the plutonium and americium isotopes inhaled in the Seascale area mainly have Class W characteristics.[2,3]

In general, the metabolic parameters used to describe the behaviour of radionuclides after their entry into the blood were those recommended by ICRP for adults[14] as there are no generally accepted age related dosimetric models. Age related models were used, however, for $^{129/131}I$, ^{14}C and ^{3}H.[17] The influence of using more appropriate age related metabolic data for calculating tissue doses was examined in a sensitivity analysis.[2] The overall conclusion was that using the adult model parameters would, if anything, tend to overestimate doses to the red bone marrow. This is mainly because of the longer retention time for caesium isotopes in adults than in young children.[18]

Radiation doses were also calculated for the developing fetus. As no dosimetric models for fetal tissues have been recommended by ICRP they had to be developed from a review of relevant animal and human data.[2] Models were developed for the most important radionuclides (isotopes of I, Sr, Cs, Pu, Am, Po).

External radiation

For exposure to external radiation from deposited activity, a dose conversion factor of 0.7 was used to convert absorbed dose rate in air, at an average height of 1 m above the ground, to absorbed dose in the tissues of the body. An overall dose reduction factor of 0.2 was used to take account of both indoor occupancy and the shielding provided by buildings.[11,19]

The direct gamma-radiation absorbed dose rates from ^{41}Ar in the passing cloud (released from the Windscale piles, up to 1957, and from the Calder Hall and the Windscale Gas-Cooled reactors) were calculated using the computer code ESCLOUD.[20] An overall dose reduction factor of 0.5 was used to allow for indoor occupancy and the shielding provided by buildings.

RISK COEFFICIENTS FOR RADIATION-INDUCED LEUKAEMIA

The risks coefficients used for fatal radiation-induced leukaemias are given in Table 2. The risk coefficients for leukaemia were based on a review of the literature.[2] The value of $3.5\ 10^{-3}\ Sv^{-1}$ used for older children makes some allowance for the improved dosimetry in the Japanese atomic bomb survivors[21] and the fact that the follow up study did not start until 1950, and therefore a substantial fraction of radiation-induced leukaemias could have been missed. The higher values used for young children and the developing fetus allow for the greater sensitivity to radiation-induced leukaemia that has

Table 2 Risk coefficients for fatal radiation-induced leukaemia

Age	Risk of leukaemia (Sv^{-1})	Period at risk (Years[c])
Fetus	$12.5\ 10^{-3}$ [a]	9
0–10 Years	$5.0\ 10^{-3}$ [b]	15
11–20 Years	$3.5\ 10^{-3}$ [b]	21

[a] 36 per cent of leukaemias assumed to occur from 0–3 years, 40 per cent from 4–6 years, 24 per cent from 7–9 years.
[b] 50 per cent of leukaemias assumed to occur in 1st third of period at risk, 33 per cent in 2nd third, remainder in last third.
[c] No latent period.
N.B. Risk coefficient recommended by ICRP[27] for adults is $2\ 10^{-3}\ Sv^{-1}$.
For high LET radiation use a Quality Factor (Q) of 20.

been observed in the fetus[22,23] and young children[24] than in adults. No latent period has been used, which is consistent with the data on leukaemia induction after radiation exposure of the fetus and young children.

RESULTS OF THE STUDY

Radiation doses to the bone marrow, and the associated risks of radiation-induced leukaemia were calculated for average children with normal habits born in Seascale at yearly intervals from 1945 to 1979. Doses and risks were calculated either to age 20 years, or to 1980 for children born after 1960, for all the main sources of radiation exposure. The risks of radiation-induced disease in the study population were based on an average birth rate of 35 children a year[25] giving a study population of 1225 children.

Figure 2 shows the average annual dose equivalent to the red bone marrow from the Sellafield discharges (other than the Windscale fire), for the average child born in 1950 (assuming Q = 20 for alpha-irradiation). This year of birth was chosen as the period of follow-up covers the entire period of interest. In the early 1950s the radiation dose to the red bone marrow was largely due to releases of ^{41}Ar from the Windscale piles. It was increased in the mid-1950s because of the release of irradiated fuel particles from the Windscale piles[26] which resulted in increased levels of radionuclides, particularly strontium and caesium isotopes, in terrestrial foodchains and external dose from deposited activity (mainly $^{95}Zr/^{95}Nb$). When the piles were shut down, following the Windscale fire in October 1957, the radiation dose to the red bone marrow progressively fell. However, there was a subsequent rise due mainly to the increasing release of radionuclides into the Irish Sea, particularly ^{137}Cs and the actinides ^{239}Pu and ^{241}Am and their entry into the marine foodchain. From 1975 onwards the discharge to the Irish Sea was reduced and this resulted in a fall in the annual doses which has continued.

Also included in Figure 2 is the radiation dose from the Windscale fire. Most of this was due either to external radiation or to intakes of ^{137}Cs and ^{210}Po, both of which have short biological half lives. As a result the dose to the bone marrow was received mainly within a year of the release.

Figure 3 shows the average annual doses to the red bone marrow, for the average child born in 1950, from the other main sources of radiation exposure. The radiation dose from natural radiation remains virtually constant; much of it arising from either cosmic rays or external gamma rays. Radiation doses

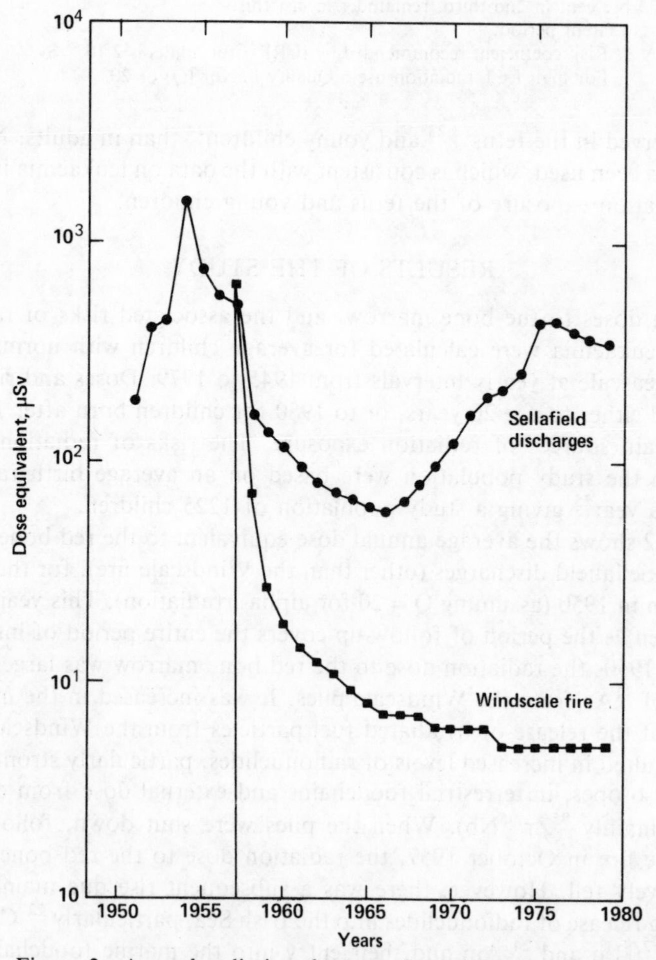

Figure 2 Annual radiation doses to the red bone marrow of the average child born in 1950 from the Sellafield discharges and the Windscale fire. Risk of leukaemia calculated to age 20 years.

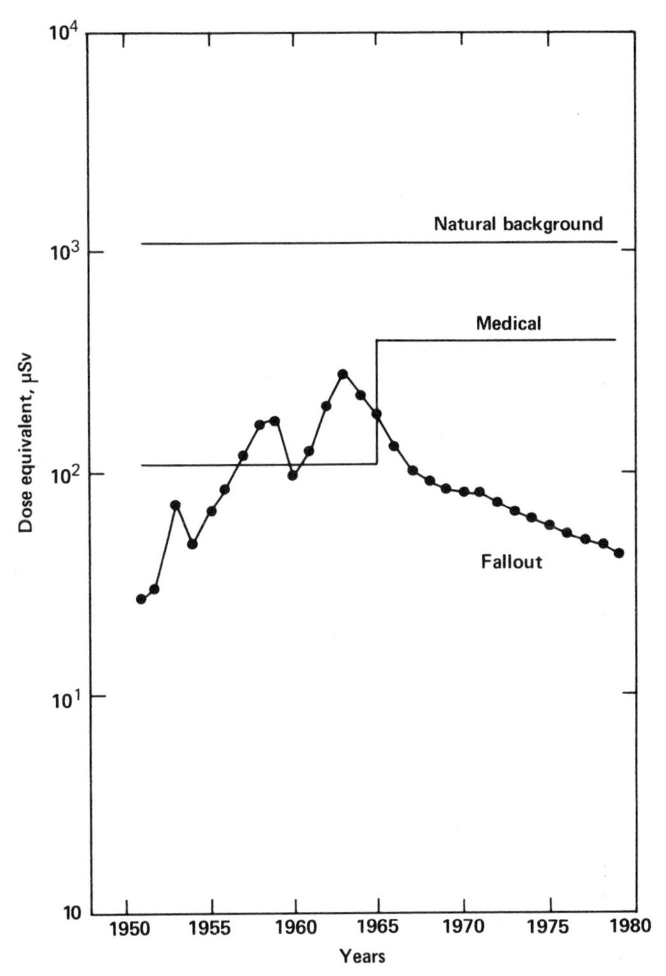

Figure 3 Annual radiation doses to the red bone marrow of the average child born in 1950 from diagnostic medical practices, nuclear weapons fallout and natural radiation. Risk of leukaemia calculated to age 20 years.

from naturally occurring radionuclides change marginally with age, but this has little effect on the average annual radiation dose.

The step increase in the radiation dose from medical exposure at age 15 years arises because only very limited information on medical exposures with age was available. Average values had to be used for the first fifteen years of life and for adults.

The radiation dose from weapons fallout follows the changing pattern of fallout deposition, reflecting the peaks in weapons testing in the late 1950s and

Table 3 Dose to the red bone marrow, by source, for Seascale children and risks of radiation-induced fatal leukaemia

Source		Date of birth 1945	1950	1955	1960	1965	1970	1975
		Dose to age 20, or 1980 (mSv),[a] and individual risk						
Sellafield discharges	Dose	4.5	5.4	5.1	3.3	3.5	3.2	1.6
	Risk	$1.5\ 10^{-5}$	$2.4\ 10^{-5}$	$2.7\ 10^{-5}$	$6.7\ 10^{-6}$	$7.5\ 10^{-6}$	$6.9\ 10^{-6}$	$3.0\ 10^{-6}$
Windscale fire	Dose	0.65	0.84	0.68	0.08	0.04	0.02	0.01
	Risk	$1.2\ 10^{-6}$	$3.6\ 10^{-6}$	$3.2\ 10^{-6}$	$3.8\ 10^{-7}$	$1.7\ 10^{-7}$	$8.0\ 10^{-8}$	$2.4\ 10^{-8}$
Weapons fallout	Dose	1.3	2.3	3.5	4.0	2.2	0.84	0.29
	Risk	$1.9\ 10^{-6}$	$6.1\ 10^{-6}$	$1.3\ 10^{-5}$	$1.8\ 10^{-5}$	$9.7\ 10^{-6}$	$2.6\ 10^{-6}$	$5.2\ 10^{-7}$
Medical	Dose	3.9	3.9	3.9	3.9	1.8	1.1	0.6
	Risk	$1.1\ 10^{-5}$	$1.1\ 10^{-5}$	$1.1\ 10^{-5}$	$1.1\ 10^{-5}$	$6.9\ 10^{-6}$	$3.3\ 10^{-6}$	$1.2\ 10^{-6}$
Natural	Dose	22	22	22	22	16	11	6
	Risk	$7.5\ 10^{-5}$	$7.5\ 10^{-5}$	$7.5\ 10^{-5}$	$7.5\ 10^{-5}$	$5.7\ 10^{-5}$	$3.6\ 10^{-5}$	$1.4\ 10^{-5}$
Total	Dose	32	35	36	33	24	17	8.5
	Risk	$1.0\ 10^{-4}$	$1.2\ 10^{-4}$	$1.3\ 10^{-4}$	$1.1\ 10^{-4}$	$8.1\ 10^{-5}$	$4.9\ 10^{-5}$	$1.8\ 10^{-5}$

[a] For calculating the dose equivalent to the red bone marrow a quality factor (Q) of 20 for high LET radiation was used.

early 1960s, and a progressive decline since then. Most of the dose arises from intakes of ^{137}Cs and ^{90}Sr and from external radiation.

The levels of radiation exposure of the population have varied with time over the period of operation of the Sellafield plant, and as a consequence the radiation dose to any child depends upon its date of birth. Table 3 gives a summary of the radiation doses to the red bone marrow, by source, for average children born every fifth year between 1945 and 1975 together with the risks of radiation-induced leukaemia. The children at greatest risk from the Sellafield discharges were those born in the mid-1950s. This was mainly as a result of discharges from the Windscale piles. The risk of radiation-induced leukaemia from the Sellafield discharges (including the Windscale fire) for children born in the mid-1950s was calculated to be about 1 in 30 000. The greatest risk from weapons fallout was to children born in the early 1960s and reflects the peak in weapons testing. For medical exposure and natural radiation the risks are constant for children born up to 1960, and then progressively fall due to the reduced period of follow up (to 1980). For all the cohorts natural radiation gave the largest radiation dose to the red bone marrow, and hence was the main contributor to the total leukaemia risk.

Table 4 summarizes the risk of radiation-induced leukaemia in the Seascale study population and the percentage contributions to this risk from the different sources of radiation exposure. A total of 0.1 radiation-induced leukaemias are predicted in the study population of 1225 children. The average risk to an individual is thus about 1 in 12 250. Natural radiation is the main contributor to the risk (66 per cent) with weapons fallout and medical exposure each contributing abut 9 per cent and the Sellafield discharges, including the Windscale fire, 16 per cent.

In the Seascale population four fatal leukaemias were observed in children up to age 20 years[1] giving an average risk of about 1 in 300. Two further

Table 4 Risk of radiation-induced leukaemia in the Seascale study population, by source

Source	Predicted number of radiation-induced fatal leukaemias in study population	Contribution to total risk (%)
Sellafield discharges	$1.4\ 10^{-2}$	14.2
Windscale fire	$1.8\ 10^{-3}$	1.8
Weapons fallout	$9.2\ 10^{-3}$	9.2
Medical	$9.1\ 10^{-3}$	9.1
Natural background	$6.6\ 10^{-2}$	65.7
Total	$1.0\ 10^{-1}$	100

children with leukaemia were alive in 1980 and one had died at age 21. From UK National Statistics 0.5 fatal leukaemias would have been expected in the population.[2] The Sellafield operations contribute about 16 per cent to the overall risk of 0.1 fatal leukaemias which corresponds to 0.016 radiation-induced leukaemias in the population. This is equivalent to an average risk to an individual child from the Sellafield discharges of about 1 in 75 000.

Within the Seascale population there will be individual children receiving both higher and lower doses than those calculated for the average child. Radiation doses were, therefore, calculated for individuals who could have received higher doses than those calculated for the average child as a result of exposure pathways that could affect only a few individuals in the population, or as a result of extreme habits.[2,3] This included, for example, children with a high consumption of sea-food, or those with the medical condition known as 'pica' in which large amounts of soil ($\sim 20\ \mathrm{g\,d^{-1}}$) are consumed. It was concluded from this sensitivity analysis that a few children in the Seascale population could have received radiation doses from the Sellafield discharges that are considerably higher than those received by the average child calculated in the main assessment. However, the increases in calculated doses, even on very pessimistic assumptions, are very much less than would be needed to account for the observed leukaemias in the population.

There remains some uncertainty about the releases from the plant and concentrations in foodstuffs in the Sellafield area, particularly for the early years of its operation. However, for the four fatal leukaemias observed in the study population to have been caused by the operations at the plant the radiation doses and risks from the Sellafield discharges would have had to be increased by a factor of about 250 for the whole period of operation of the plant. The possibility that the average doses calculated in this study and the risk coefficients used for radiation-induced leukaemia could be so substantially wrong is most unlikely.

SUMMARY AND CONCLUSIONS

This chapter has reviewed an assessment by the National Radiological Protection Board of the risks of radiation-induced leukaemia in children and young persons in Seascale over the period of operation of the Sellafield nuclear fuel reprocessing plant up to 1980. The analysis has shown that in the study population of 1225 children born in the village between 1945 and 1979, and followed to 1980, 0.1 radiation-induced leukaemias would be expected in children under 20 years of age from all radiation sources. This corresponds to a risk of about 1 in 12 250 for the average child in the study population. Natural radiation contributes about two-thirds of this risk and the Sellafield discharges (including the Windscale fire) 16 per cent. Nuclear weapons fallout and medical exposure each contribute about 9 per cent. The risk to the average child in the study population from the Sellafield discharges is calculated to be about 1 in 75 000, with a maximum risk of about 1 in 30 000 for children born in the mid-1950s.

REFERENCES

1. D. Black, *Investigation of the Possible Increased Incidence of Cancer in West Cumbria*. Report of the Independent Advisory Group. Chairman Sir Douglas Black. HMSO, London, 1984.
2. J. W. Stather, A. D. Wrixon and J. R. Simmonds, The risks of leukaemia and other cancers in Seascale from radiation exposure. *NRPB-R171*. HMSO, London, 1984.
3. J. W. Stather, J. Dionian, J. Brown, T. P. Fell, and C. R. Muirhead, The risks of leukaemia and other cancers in Seascale from radiation exposure. *NRPB-R171-Addendum*. HMSO, London, 1986.
4. J. R. Simmonds, G. S. Linsley and J. A. Jones, A general model for the transfer of radioactive materials in terrestrial foodchains. *NRPB-R89*. HMSO, London, 1979.
5. J. R. Simmonds and M. J. Crick, Transfer parameters for use in terrestrial food-chain models. *NRPB Memorandum M63* (1982).
6. G. J. Hunt, Assessment of public radiation exposure due to liquid effluents from fuel reprocessing. Part 1; Individual (critical group) dose. *Sizewell Inquiry Series No. 8, MAFF*/S/II (Saf), 1982.
7. MAFF (Ministry of Agriculture, Fisheries and Food), *National Food Survey Committee, Household Food Consumption and Expenditure; 1981*. Annual report. HMSO, London, 1983.
8. MAFF (Ministry of Agriculture, Fisheries and Food), Annual reports on radio-activity in surface and coastal waters of the British Isles, 1966–1981 (1967–1983).
9. A. D. Wrixon, L. Brown, K. D. Cliff, C. M. H. Driscoll, B. M. R. Green and J. C. H. Miles, Indoor radiation survey in the UK, *Radiat. Prot. Dosim.*, **7**, 321 (1984).
10. UNSCEAR, *Sources and Effects of Ionizing Radiation*. United Nations Scientific Committee on the Effects of Atomic Radiation 1977 Report to the General Assembly, with Annexes. UN, New York, 1977.
11. UNSCEAR, *Ionising Radiations: Sources and Biological Effects*. United Nations Scientific Committee on the Effects of Atomic Radiation 1982 Report to the General Assembly, with Annexes. UN, New York, 1982.
12. Adrian, Radiological hazards to patients; Final report to the committee under Lord Adrian. HMSO, London, (1966).
13. G. M. Kendall, S. C. Darby, S. V. Harries and S. A. Rae, (1980). A frequency survey of radiological examinations carried out in National Health Hospitals in Great Britain in 1977 for diagnostic purposes. *NRPB-R104*. HMSO, London, 1980.
14. ICRP, Limits for intakes of radionuclides by workers. *ICRP Publication 30, Annals of the ICRP, Volume 2, No. 3/4*. Pergamon Press, Oxford, 1979.
15. N. Adams, Dependence on age at intake of committed dose equivalents from radionuclides, *Phy. Med. Biol.*, **26**, 1019 (1981).
16. ICRP, The metabolism of plutonium and related elements: report of a Task Group of the International Commission on Radiological Protection. *ICRP Publication 48, Annals of the ICRP, Volume 16, No. 2/3*. Pergamon Press, Oxford, 1986.

17. J. R. Greenhalgh, T. P. Fell and N. Adams, Doses from intakes of radionuclides by adults and young children. *NRPB-R162*. HMSO, London, 1985.
18. M. A. Cryer and K .F. Baverstock, Biological half-life of ^{137}Cs in man, *Health Phy.*, **23**, 394 (1972).
19. L. Brown, National radiation survey in the UK, indoor occupancy factors, *Radiat. Prot. Dosim.*, **5**, 203 (1983).
20. J. A. Jones, ESCLOUD: A computer program to calculate the air concentration, deposition rate and external dose rate from a continuous discharge of radioactive materials to atmosphere. *NRPB-R101*. HMSO, London, 1980.
21. W. E. Loewe and E. Mendelsohn, Revised dose estimates at Hiroshima and Nagasaki, *Health Phys.*, **41**, 663, (1981).
22. A. M. Stewart and G. W. Kneale, Radiation dose effects in relation to obstetric X-rays and childhood cancers, *Lancet*, **1**, 1185 (1970).
23. R. R. Monson and B. MacMahon. 'Prenatal X-ray exposure and cancer in children, in *Radiation Carcinogenesis, Epidemiology and Biological Significance*. (Ed. J. D. Boice *et al.*). Raven Press, New York, (1984).
24. G. Beebe, H. Kato and C. E. Lane, Studies of the mortality of A-bomb survivors. Mortality and radiation dose 1950–74, *Radiat. Res.*, **75**, 138 (1978).
25. E. Rubery, Department of Health and Social Security, London, Private communication, 1984.
26. ICRP, Report of the Task Group on Reference Man. *ICRP Publication 23*. Pergamon Press, Oxford, 1974.
27. ICRP, Recommendations of the International Commission on Radiological Protection. *ICRP Publication 26, Ann. of the ICRP, Volume 1*, No. 3. Pergamon Press, Oxford, 1977.

Radiation and Health
Edited by R. Russell Jones and R. Southwood

6

Discussion Period 1

Dr Blomfield (*general practitioner, Yorkshire*) Frances Fry mentioned external monitoring of radiation in the body, as a way of assessing intake of radiation? But what about alpha-radiation.

Ms Fry The only thing you can measure are those gamma rays that actually escape from the human body.

Dr Blomfield So the alpha and beta emissions cannot be monitored internally?

Ms Fry No, but urinalysis is possible, as is done in occupational exposure.

Dr Blomfield Surely that depends on how much is excreted.

Ms Fry Yes, but we still need models to relate what is measured, to doses received, no matter what we measure.

Dr Blomfield Yes, but I understand you to say that your methods provide the most accurate way of measuring the internal uptake of radionuclides.

Ms Fry For the radionuclides we can measure, yes.

Dr Blomfield But alpha and beta are surely the most important, and you cannot assess these using this method. In life that is.

Ms Fry No, but I think a great deal of exposure comes from radionuclides that do emit gamma rays. I admit that we would like to be able to measure the others, but we cannot easily do so. Doses have to be estimated from environmental measurements or indeed from bioassay procedures.

Dr Blomfield I would suggest that the only way you can really assess the problem is post-mortem.

Ms Fry That is certainly a useful technique.

Dr Blomfield It is the only accurate technique.

Ms Fry Accurate as a measure of what is in the body at the end of life, yes. One still needs to relate that to what has gone on during the person's lifetime.

In other words you are measuring the end result. You still need to know how much activity that person has taken in, on an annual basis.

Dr Blomfield I would suggest that if I have taken in significant amounts of strontium-90, wherever it is come from, I could not be assessed accurately even if it is important in the induction of leukaemia.

Ms Fry I think the radionuclides that are important are ones like strontium-90 and plutonium and they are the ones that are difficult to assess.

Dr Blomfield Or impossible to assess?

Ms Fry We certainly cannot measure them directly in the body. We have to use these other techniques.

Mr David Lowry (*Energy Research Group, Open University*) Last Thursday the House of Lords had a debate on nuclear power in Europe and the 28 speakers endorsed an expansion of nuclear power in Europe and in Britain in particular. Many of their Lordships spoke about the misunderstanding of the risks of radiation, making the point that most members of the public thought radiation is a lot more dangerous than it is. One of their Lordships said that you should compare the dangers of discharges from coal-fired stations to nuclear stations and he said,

'The smoke from power stations can be cleaned up and the acid removed, but it is expensive; even that is not the end of the problem. The residue left is very considerable and ironically contains not less but more radioactivity than the waste from nuclear power which causes so much public concern.'

The question I would like to put to Frances Fry of the NRPB is: could you give me a comparative assessment of the amount of radioactivity in residues from coal-fired stations, compared with that contained in nuclear waste, assuming he means by nuclear waste, spent fuel and not just waste after reprocessing?

Ms Fry I am sorry I am not sure that I can answer your question in quantitative terms. I believe it right that there is more radioactivity discharged from coal-fired than from nuclear power stations.

Dr Klarissa Nienhuys (*Dutch Society for the Preservation of the Waddensea*) As far as I know, only very dirty coal plants, which would not be allowed to be built in any decent Western country could compare in terms of release, of radionuclides, with a nuclear power station. Emissions from nuclear power stations are much higher, usually by a factor of 10.

Dr Philip Little *Safety Officer, UK Atomic Energy Authority, Culham Laboratory*) Friends of the Earth in evidence to the House of Lords Select Committee, following the Chernobyl accident, stated that if one takes the whole nuclear industry—from mining, transport, use through to final disposal—and compares it with the coal industry, the risk to the public from those two forms of energy supply is equivalent, and that takes into account the possibility of catastrophe and long-term effects on the population.

Mr Stewart Boyle (*Friends of the Earth*) In fact our evidence to the House of Lords, comparing the nuclear and the coal fuel cycles, was under normal operation and it expressly excluded catastrophic events. The point we were making was that under normal operation there is not much difference between both cycles. The factor which differentiates the nuclear fuel cycle, of course, is the low probability but massive consequence of a catastrophic accident.

Dr Robin Russell Jones (*Conference Organizer*) My question is directed to Dr Stather as the author of the report that produced all the controversy at Sellafield. I believe the original report was almost 300 pages long and you produced an addendum this year which was 158 pages long, a very prodigious effort. We have heard something of the uncertainties that are contained within these pages. But the crucial question that you were trying to answer is what the actual dose is to the cell that transforms and produces a lymphatic leukaemia, because most of the leukaemias that were observed around Sellafield and Dounreay were of the lymphatic variety. And yet in your addendum, on page 38, you actually say, 'There is no definitive information on the sites of the sensitive cells involved in the production of lymphoid tumours in man. The target cells for malignant transformation in the lymphoid leukaemias are unknown'. How is it possible to produce a risk estimate—sometimes to the nearest decimal point—when you are not certain where the cell is that you are trying to calculate the dose to?

Dr John Stather As I indicated I would not put a lot of belief in the second decimal point. Maybe there are uncertainties within a factor of two or three, but not two to three order of the magnitude which is what is needed. Recently Professor Greaves from the Institute of Cancer Research has stated that the committed lymphoid precursors of the 'B' series, many of which are found in the bone marrow, are probably the target cells for lymphatic leukaemia. So in that case we would in fact be calculating the dose to he right tissue.

Dr Russell Jones Except of course that lymphocytes are incredibly mobile cells. They move around the body from lymph node to bone marrow and there are as many uncertainties about the dose to lymph node, as they are doses to the bone marrow. In addition, you are dealing with alpha-emitting radionuclides such as plutonium and americium which are not present in nature. Therefore your assumptions about the proportional dose from artificial and background origin may be wrong.

Dr Stather That seems to imply that cells are going to respond differently to radiations from artificial nuclides than to those from naturally occurring ones which I would doubt. Also, in both cases—natural radiation and the Sellafield discharges—alpha radiation contributes about 10 per cent of total dose. So there would be a greater alpha dose from natural radiation than from the Sellafield discharges, at least for the average case.

Professor R. Scorer (*Imperial College*) I do not understand why Frances Fry and Dr Stather are always insisting on calculating the average dose distributed over the whole population, when we are actually dealing with a very small minority. What you should be saying is surely, can we think of any pathway by which these few people could actually receive the necessary dose to get leukaemia? I do not see why you should have to require the whole population to have to suffer a certain dose in order that this tiny minority should get leukaemia. Now maybe this is a statistical trick, well I think it *is* a statistical device of the people who use statistics, in order to avoid having to think about the real mechanics of what is actually going on.

Dr Stather We have actually looked at doses to individuals as well, who may be more highly exposed. The difficulty is, if you are going to say an individual child is so highly exposed that he stands a one in one, or even one in ten chance, of developing leukaemia, the radiation dose of that child would have to be incredibly high, enough to produce non-stochastic effects.

Professor Scorer Therefore, before you have done any calculations, you are saying it is impossible for the leukaemia to result from the discharges. Your model does not allow for that possibility.

Dr Stather We have looked at what possible doses result from individual children with extreme eating habits. The child who eats a large amount of seafood, or sediment, will receive doses maybe twenty to thirty times those that we calculate for the average child. But even that is not a dose approaching what would need to be received, to cause the leukaemias that have been observed. These doses would, on the basis of our risk estimates, have to be 250 times greater for the whole study population.

Ms Ann Link (*ALARM*) I would like to ask the panel their views about the constant comparison of artificial nuclides with the natural ones.

Mr David Crouch The argument used is that if the artificial discharges were responsible for the leukaemias then we would observe an absurdly large number of cancers from background radiation. There are two points to be made here. First, there has been innumerable surveys trying to correlate background radiation with cancer incidence. As one reviewer remarked, the number of factors that can intrude to vitiate any such analysis makes it surprising that so many of these analyses have indeed found a positive correlation. Second, we have to consider the dose response to alpha-emitters. The alpha dose to bone marrow from natural radiation is quite small; so we could envisage a much higher risk factor for alpha radiation without it necessarily conflicting with background cancer incidence.

Part 3

Cancer Risk

Radiation and Health
Edited by R. Russell Jones and R. Southwood

7

Recent Evidence of Radiation-induced Cancer in the Japanese Atomic Bomb Survivors

EDWARD P. RADFORD MD
Adjunct Professor of Epidemiology
University of Pittsburgh, Pennsylvania, USA

ABSTRACT

The prospective study of survivors of the atomic bombing of Hiroshima and Nagasaki utilizes as a control population those persons who were far enough from the detonation to receive no exposure, but who were in the cities at the time of the bombing (ATB). Radiation exposure data are still being revised; the new evaluation indicates that the results from the two cities can be combined, since fast neutron exposures were low and were similar in both cities. New tissue doses are about half those presented previously. Many confounding factors affecting cancer rates have also been evaluated, such as medical radiation, smoking, childbearing, diet and indoor radon. The most recent follow-up data support the following conclusions: (a) the dose–response relationship is consistent with a straight line through the origin, including the lowest dose group (~ 3 rad); (b) sensitivity to induction of cancer varies considerably by tissue irradiated; (c) most cancers show a radiation effect still increasing 40 years after exposure; (d) a small leukaemia excess among those irradiated is still present in Hiroshima; (e) the thyroid cancer excess is declining at this time; (f) smoking adds to the effect of radiation on lung cancer incidence; (g) certain benign tumours show a radiation-related effect; (h) children under the age of 10 ATB are presently showing the highest relative risk for cancer compared with all other ages ATB at equal attained age. If this last effect continues to persist, then age-specific lifetime cancer risk coefficients will be necessary, and for those irradiated as young children may be quite high.

The follow-up studies of survivors of the atomic bombing in Hiroshima and Nagasaki have now extended for 40 years, the largest prospective human epidemiologic study yet undertaken. In defining the effects of low doses of ionizing radiation, the A-bomb survivor study is especially important because it involves a group of people of all ages and both sexes who were in reasonable health prior to radiation exposure. Among the survivors identified in 1950 and who are included in the study population the average tissue dose for those exposed to 1 rad or more is about 20 rad, and thus many results are especially relevant to effects of low doses of radiation, contrary to the usual view that the A-bomb findings are primarily related to high doses.

Until recently it was thought that the type of radiation exposure in the two cities differed significantly, in that the Hiroshima uranium-235 bomb was believed to have a much higher proportion of high energy neutron exposure at ground level than the Nagasaki plutonium-239 bomb. The most recent reassessment of dosimetry,[1,2] however, has indicated that the distribution of high-energy neutrons by distance from ground zero was very nearly the same in both cities. Although the free-in-air gamma-ray doses continue to be proportionately somewhat higher in Nagasaki compared with Hiroshima at equivalent distances, the neutron component in both cities is now too small to provide any conclusive basis for ascribing differences between results in each city to the quality of radiation exposure.

In addition the shielding of gamma rays by houses and other buildings, in which most of the survivors were present at the time of exposure, was formerly under-estimated; the new dosimetry system leads to a reduction of doses (about twofold) because of the reduction of neutron dose and the greater shielding of gamma rays. With this system applied, the dose–response relationship for radiation-induced cancer is very similar in the two cities (within about 3 per cent). The results for both cities can be combined, therefore, with an improvement in the statistical reliability of the resulting cancer risk estimates.

The population under study at the Radiation Effects Research Foundation (RERF) for assessment of long-term effects of ionizing radiation is the Life Span Study (LSS) sample; follow-up of this sample began in October 1950. The LSS sample consists of persons who were living in either Hiroshima or Nagasaki at the time of the census of A-bomb survivors conducted by the Japanese government in 1950. As originally defined, the sample included (1) most persons alive at that time who were within 2500 m from ground zero in either city at the time of the bomb (ATB), (2) a sample of persons who were between 2500 and 10 000 m from ground zero ATB, and (3) a sample of persons who were not in city (NIC) or beyond 10 000 m. The latter two samples were obtained by matching by sex and age to a group of survivors who were less than 2000 m from ground zero in each city. The original sample has been extended, most recently in 1982 by the addition of approximately 11 400 distally exposed (2500–10 000 m) Nagasaki residents for whom complete follow-up data are available. As currently defined and with the NIC group eliminated, the total study population is presently 93 614 (61 911 in Hiroshima and 31 703 in Nagasaki); about one-third have died in the period 1950 to 1982. Analysis of the data has used those who were about 3000 m or more from ground zero (and thus whose doses were less than 0.5 rad) as the control (non-exposed) population.

An important subset of the LSS is the Adult Health Study (AHS) population, which had had extensive and continuing clinical evaluations beginning in 1958. This group was selected to include a high proportion of individuals exposed to high doses, and a random sample of survivors with lower doses. In 1958 a total of 19 962 persons were selected from the entire LSS population, as it was then defined. Of these, 16 738 had been examined at least once by the end of 1978. In 1977, another 2436 persons were added to the AHS population, and over 60 per cent of these were examined at least once within four years. All AHS subjects living in or near either city are encouraged to return every two years for detailed evaluation of their health status, including medical history, physical examination, laboratory evaluations, and special clinical studies as needed. This population has been useful in detecting radiation dose-related effects on the prevalence of benign tumours.

Mortality data within the LSS sample are obtained by periodic checks of the Japanese system of local family registration offices (*Honseki*). These offices maintain records of family vital events, births, marriages, deaths, etc., and each living sample member's record (*Koseki*) is periodically reviewed to determine vital status. Whenever a record of a person's death is found, the date and cause of death are obtained from the vital statistics death schedule at the Health Centre for the place of death; the cause of death reported by these centres is obtained from death certificates. Because of the comprehensive nature of the Koseki system, ascertainment of death is believed to be virtually complete for subjects resident in Japan. The hazards of relying on death certificates to establish the primary cause of death are well known. A comparison of certified cause of death and autopsy diagnoses[3] revealed that cancer confirmation and detection rates vary widely according to the type of cancer. The accuracy of death certificates in the LSS population was found to be high for malignant neoplasms of some organs such as breast and stomach, and for leukaemia. But for others, such as lung, urinary tract, liver and biliary system, pancreas and prostate, death certificate data were found to underestimate the presence of cancer considerably.[4] This under-estimation will produce an equivalent under-estimation of the absolute excess risks associated with exposure to radiation.

Information on cancer incidence is also obtained through the tumour registries established in both cities by the local medical societies in the late 1950s, tissue registries in each city, and supplementary case finding efforts conducted for particular studies. At this time reporting of cases by the tumour registries is considered to be good for the two cites. The registries incorporate cases only for those resident in each city at the time of the diagnosis, thus there is under-ascertainment of cases in the LSS population because of migration out of the cities, a particular problem for those who were young ATB. Data from the AHS population indicate that since 1958 out-migration has been slight (only a few per cent). However, prior to that time about 10 per cent of all those living had left. As many as 20 per cent of the members of the youngest age ATB groups may have left the cities prior to 1958. Tokunaga *et al.*[5] cited evidence that among AHS females the migration rate was unrelated to radiation dose. Because most of those leaving have settled in the large cities of Honshu and Kyushu, it is possible that they could be traced to tumour registries there, in order to bring the ascertainment of incident cancer closer to 100 per cent.

The cancer incidence data have the advantage that cases are included in the study at the time of diagnosis, often years before fatal cancer would be found from death certificate records. Not only may there be a substantial lag between diagnosis and death, but sometimes as much as two years may elapse before the death certificate data are recorded and detected from the Koseki. The delay in ascertainment from death records is especially important because cancer cases in the LSS are rapidly accumulating, and the evidence of radiation dose-related effects is increasing.

Another important factor in use of tumour registry data is the fact that cases are recorded regardless of outcome. Thus information on non-fatal malignancies and some types of benign tumours can be obtained from the registries. This issue is especially significant for cancer of the thyroid and female breast, which are radiosensitive but not highly fatal. The radiosensitivity of cancer of the prostate is uncertain, in part because ascertainment of cases of cancer of the prostate from death certificates is poor in Japan. Data from the tumour registries, for which ascertainment is somewhat better, indicate a radiation dose-related effect.[6] Finally, it should be emphasized that use of incident cancer as a basis of defining cancer risks gives an indication of radiation effects that is generally more meaningful in terms of total social cost than simply cancer deaths alone. Incident cases are not yet completely analysed for the period up to 1982, and therefore the main emphasis in this report will be on mortality evaluation.

Host and environmental characteristics such as smoking habits, diet, socioeconomic status, childbearing history, occupation, and many other details have been collected in a series of interview and mail surveys of large subsets of the LSS sample conducted at various times between 1963 and 1981. These data have been useful in efforts to deal with factors other than A-bomb radiation exposure which might affect cancer risks. An important potential confounding factor is exposure of the study population to medical X-rays. Evaluation of this source of additional radiation exposure could be important in future analyses of LSS data, particularly for those subjects given X-ray therapy for benign conditions. A survey of hospitals in the Hiroshima area, as well as among the AHS population, indicates that average cumulative radiation exposure to the bone marrow or gonads since 1945 has been less than 2 rad from diagnostic X-ray.[7] With regard to therapeutic radiation, local doses in excess of 1000 rad have been received by approximately 2000 persons in the LSS.[8] About 65 per cent of this therapy has been for cancer, therefore among this group the primary interest is in the possibility of second primary malignancies.

RECENT RESULTS

A preliminary analysis of follow-up of mortality through 1982 has been carried out using the most recent estimates of revised dose data available in 1986. Figure 1 shows the dose–response for all malignant neoplasms except leukaemia, 1955–82, with results for both cities combined. Risk relative to the zero dose group is plotted against mean tissue dose. In calculating the dose a quality factor of 10 for the neutron component has been applied, and tissue

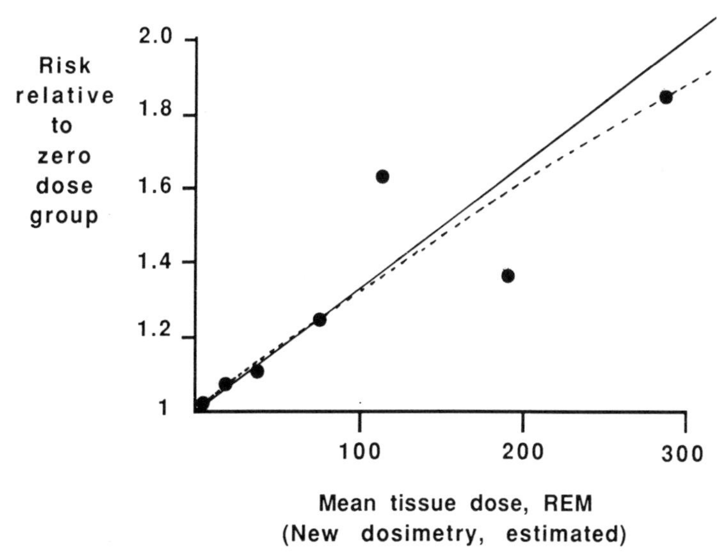

Figure 1 Cancer mortality in LSS sample plotted against mean tissue dose in rem, for all malignant cancers except leukaemia. Data for the period 1955 to 1982, both cities and both sexes combined, and all ages. Ordinate: Cancer risk relative to the unexposed group. Doses estimated from new dosimetric system, with a neutron quality factor of 10 applied. Solid line: straight line fit to data drawn through origin. Dashed line: line fitted to equation. Effect = $aD + bD^2$ where D is dose and a and b are constants. The downward curvature indicates that the coefficient b is negative.

doses have been calculated assuming a constant value of 0.7 for tissue transmission.[9] Grouping by dose categories has been on the basis of the old dosimetry, corrected for the new changes at each level, but there will be some reassignment of individuals on the basis of the new dosimetry. Thus these results may be modified somewhat by further analysis. The period from 1950 to 1954 has been excluded because except for leukaemia the person-years 5–10 years from the exposure are at low risk from radiation-induced cancers due to the minimum latent period required for the development of the solid cancers.

A number of points may be made from Figure 1. First, for the four lower doses, at least, the relationship is reasonably close to a straight line down to and including the dose at 3 rem. At higher doses there is evidence that people may have been misclassified by dose; these were people in whom shielding by building construction or materials may have determined their survival, and for whom shielding estimation is less accurate. If an equation of the form $E = aD + bD^2$ is used to fit the data points, where E is the excess relative cancer risk, D is dose, and a and b are constants, the coefficient b is found to be negative. In other words, the best fit of the equation in this case is curvilinear downward, implying greater effect per unit dose at low doses than at high doses. The two lowest dose points, at 3 and 16 rem, have 2606 out of 3383 cases observed, or 77 per cent of all cancer deaths among those observed in all groups exposed to 1 rem or more. Moreover these groups, exposed at greater distance from ground zero, have more reliable dosimetry, which is less sensitive to errors of location and shielding, than the groups exposed to higher doses. As of 1982, the lowest dose group shows a 3.26 per cent excess relative risk of cancer for a mean tissue dose of about 3 rem, or a doubling dose for all cancers except leukaemia (dose required to double the cancer mortality) of 92 rem. For the second data point the excess relative risk is 9.14 per cent for a mean tissue dose of about 16 rems or a doubling dose of about 175 rem. If we pool the results from these two data points, with a combined average dose (weighted) of about 7.4 rem, the doubling dose is about 138 rem. For the pooled results for the remaining five higher dose categories, with a weighted mean dose of 94.0 rem (777 total cancer deaths observed), the doubling dose is about 315 rem. The lower doubling dose at the low doses is an expression of the downward curvature of the cancer risk dose–response derived from these data.

When one looks at individual sites of cancer among the A-bomb survivors, the radiation dose-related excess risk relative to the zero dose group is quite variable, indicating that some types of cancer may be more easily induced by radiation than others. Some of this variability in the cancer mortality results may depend on inaccuracies of death certification, especially a problem in Japan for cancers of the prostate, uterine cervix, pancreas and liver and biliary system.[4] For example *deaths* certified to prostate or pancreatic cancers show no relationship with radiation dose, up to 1982, but a positive dose–response relationship has been demonstrated for prostatic and pancreatic cancers obtained from the tumour registries. Despite these questions it is evident that thyroid cancer (incidence), leukaemia, cancer of the female breast, urinary tract cancer, lung cancer and multiple myeloma are among the most sensitive to radiation in the A-bomb survivors. One should note that the ability to detect these effects in the LSS sample depends on the cancer frequency; for cancers diagnosed rarely in Japanese, such as bone cancer or Hodgkin's disease, no radiation effect may be detectable because the numbers observed are too small.

Figure 2 shows the dose–response for leukaemia deaths in the LSS sample for the period 1950 to 1982. There are only 158 deaths distributed among the seven dose categories, thus the statistical reliability of each point is much less than for the points in Figure 1. Nevertheless the dose–response apparently shows a curvilinear relationship somewhat different than is found in Figure 1, with low doses less effective per unit dose than the high doses. One should note, however, that the relative risks on the ordinate of Figure 2 are very high.

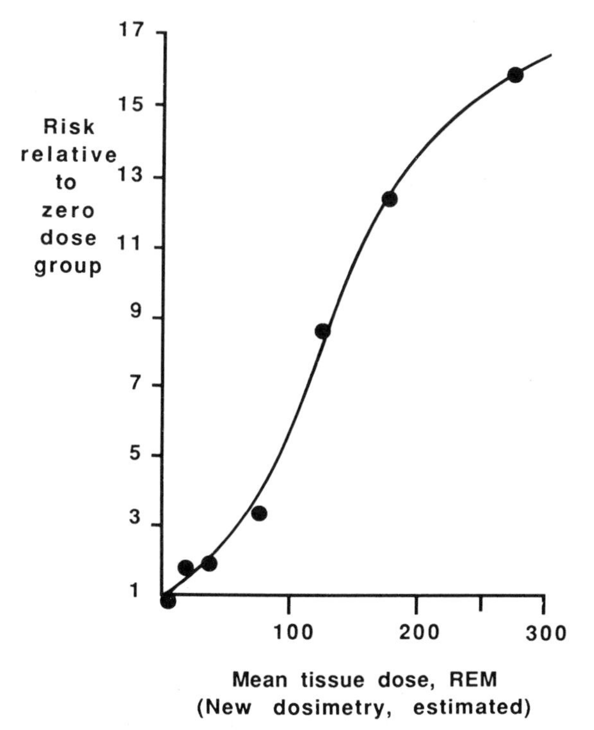

Figure 2. Leukaemia deaths in LSS sample versus mean tissue dose in rem. Data for the period 1950–82, both cities and both sexes combined, and all ages. Ordinate and abscissa as in Figure 1. The solid line is fitted by eye.

If a straight line is drawn through the four lowest dose points and the origin, the doubling dose is found to be about 35 rems for all age groups combined.

Investigation of the trend of relative risk by four-year time intervals since 1950 shows that the excess risk of leukaemia per rem has declined steadily since 1958, although a relative risk of about 2 (5 cases versus 2.7 expected) still was observed in the period of 1979–82 for those exposed to greater than about 20 rem tissue dose. In contrast, for all cancers except leukaemia the excess relative risk per rem for the exposed population is continuing to increase during the period 1959 to 1982, and is highest for the period 1975 to 1982. Overall there is no evidence for cancer mortality of a decline in excess relative risk up to 1982, 37 years after the exposure. For some particular cancers besides leukaemia, however, a decline in risk with time appears to be present. Thyroid cancer incidence per rem is beginning to decline, and so also is multiple myeloma, although the number of cases, 35, is too small to be certain.

The excess relative risk per rem of all cancers except leukaemia is about twice as high in women compared with men. This sex difference is largely due to the low cancer age-specific death rates among women compared with men. In absolute terms there is little difference by sex in excess cancer mortality per rem, though it is likely to be present in analyses of cancer incidence because of the importance of breast and thyroid cancer (with high survival rates) among women. For leukaemia there is no sex difference in relative risk per rem, but in absolute terms the risk is about twice as high in males.

An important finding with continuing follow-up of this population is the effect of age at exposure on subsequent cancer risk. Those survivors who were under age 10 ATB are now reaching ages when cancer is beginning to occur more frequently, and thus it is beginning to be possible to determine their excess risk at the same attained age as those who were older ATB. When this is done it is evident that the excess relative risk is generally substantially higher for those irradiated at young ages. As of 1982 the excess relative risk per rem for those children under 10 years of age ATB is about eight times higher for all cancers except leukaemia, compared with those over age 35 ATB. This difference in effects of age at exposure may decline as the cancer rates rise with older age in this youngest group, but if it is found that the higher excess relative risk in this group persists throughout their lives, then it is apparent that children are at special risk of developing cancer from radiation exposure. In the case of leukaemia, for which the excess risk is now nearly completely defined for all age groups, the excess relative risk for those under age 10 ATB is about four times greater than for those over 35 ATB. With all dose categories combined, the doubling dose for leukaemia is about 8 rem for children exposed under the age of 10. A similar effect of age at exposure has been demonstrated for the incidence of breast cancer in women exposed under the age of 10.[5]

With regard to the additional personal or environmental factors which could interact with the radiation-related cancer risk (confounding by other randomly-present contributors to cancer would be expected to reduce the apparent effect of radiation), most have not been found to modify the radiation effects significantly. In the case of cigarette smoking in relation to lung cancer, a case-control evaluation of lung cancer cases has indicated that the effects of smoking and radiation were nearly additive together. Thus, among women, most of whom did not smoke, the absolute excess of lung cancer related to radiation exposure was about the same as for men, most of whom did smoke. Unexposed women had a low rate of lung cancer, thus the excess *relative* risk from radiation exposure was much higher than for men, whose lung cancer rates were high because of smoking, among those unexposed to radiation.

The mortality analysis does not adequately detect benign tumours because they usually are not recorded as the primary cause of death. It has been possible to determine benign tumours in the AHS sample examined every two years in the RERF clinic. Conditions which show a dose-related effect of radiation in this population include gastric polyps, uterine fibromas and non-malignant thyroid disease.[10] The possibility of a relationship of these benign conditions to subsequent development of malignant cancer at these sites is of interest.

In summary, the follow-up study of the A-bomb survivors continues to yield important results. The new evidence indicates that the risk per rem of radiation-induced cancer is higher than previously thought for the following reasons:

(a) The new evaluation of radiation exposures indicates that the doses were previously overestimated, thus it is now evident that the excess cancers were produced with lower radiation doses than were thought to apply;
(b) For most cancers the excess relative risk in those irradiated is continuing to increase with time of follow-up;
(c) Those irradiated at young ages are showing higher relative cancer risks per rem than those who were older at the time of exposure, when compared at the same attained ages. The straight line relationship between radiation dose and excess cancer risk is well-supported for all cancers except leukaemia. The data suggest that at low doses the relative risk is an increase in all cancers including leukaemia of about 0.75 per cent per rem for this exposure condition, consistent with a doubling dose of 133 rem. It is likely that this relative risk coefficient will be found to depend on age at irradiation, with an excess cancer risk of 2 per cent per rem or more for those irradiated under the age of 10. Based on these results and the normal spontaneous cancer rates in Japan (only about 60 per cent as high as those

found in Western countries), the lifetime absolute excess risk is at least 1 excess (incident) cancer case per 1000 persons exposed per rem, for all ages, and substantially higher for irradiated children.

ACKNOWLEDGEMENTS

Portions of the section on methods have been derived from E. P. Radford, D. Preston and K. J. Kopecky, Methods for study of delayed health effects of A-bomb radiation, *GANN Monograph on Cancer Research*, **32**: 75–87, 1986. I am indebted to Drs Dale Preston and Kenneth J. Kopecky for providing unpublished data on which this chapter is based. The statements and conclusions in this chapter are those of the author only and do not represent those of any organization.

REFERENCES

1. G. D. Kerr, J. V. Pace III and W. H. Scott, Jr. 'Tissue kerma vs. distance relationships for initial nuclear radiation from the atomic bombs, Hiroshima and Nagasaki' (1983). In *Reassessment of Atomic Bomb Dosimetry*, pp. 57–103. RERF, Hiroshima, 1983.
2. W. A. Woolson, M .L. Gritzner and S. D. Egbert, *Atomic Bomb Survivor Dosimetry*. Report to RERF, Dec. 1984.
3. A. Steer, I. M. Moriyama and K. Shimizu, The autopsy program and the life span study, January 1951–December, 1970. *ABCC Tech. Rep. 16-73* (1973).
4. E. P. Radford, A comparison of incidence and mortality as a basis for determining risks from environmental agents, in *Proc. of the 20th Annual Meeting of the National Council on Radiation Protection and Measurements*, pp. 75–88, NCRP, Bethesda, Md. (1985).
5. M. Tokunaga, C. E. Land, T. Yamamoto, M. Asano, S. Tokuoka, H. Ezaki and I. Nishimori, Incidence of female breast cancer among atomic bomb survivors, Hiroshima and Nagasaki, 1950–80. *RERF Tech. Rep. 15-84* (1984).
6. T. Wakabayashi, H. Kato, T. Ikeda and W. J. Schull, Studies of the mortality of A-bomb survivors, Report 7. Part III. Incidence of cancer in 1959–78, based on the Tumor Registry data, Nagasaki, *Radiat. Res.*, **93**, 112–46 (1983).
7. W. Sawada, C. E. Land, M. Otake, W. J. Russell, K. Takeshita, H. Yoshinaga and Z. Hombo, Hospital and clinic survey estimates of medical x-ray exposures in Hiroshima and Nagasaki. Part I, RERF population and the general population. *RERF Tech. Rep. 16-79* (1979).
8. J. A. Pinkston, S. Antoku and W. J. Russell, Malignant neoplasms among atomic bomb survivors following radiation therapy. *Acta Radiol. Oncol.*, **20**, 267–71 (1981). (Also personal communication, W. J. Russell, 1984.)
9. G. D. Kerr, K. F. Eckerman, J. S. Tang, J. C. Ryman and M. Cristy, Organ dosimetry, in *Reassessment of Atomic Bomb Dosimetry*, pp. 79–82, RERF, Hiroshima, (1983).
10. H. Sawada, K. Kodama, Y. Shimizu and H. Kato, RERF Adult Health Study Report 6: Results of six examination cycles, 1968–1980. Hiroshima and Nagasaki. *RERF Tech. Rep.*, in preparation.

Radiation and Health
Edited by R. Russell Jones and R. Southwood

8

Epidemiological Studies of Workers in the Nuclear Industry

VALERIE BERAL, MRCP, PATRICIA FRASER, MD, MARGARET BOOTH, PhD and LUCY CARPENTER, MSc
Epidemiological Monitoring Unit, Department of Epidemiology, London School of Hygiene and Tropical Medicine, London, UK.

Eight follow-up studies, together covering the experience of 120 000 workers in the nuclear industry, have now been published. In all the workforces the overall mortality rates were below their respective national rates, this being consistent with the relatively high social classes of the employees and the initial selection of healthy persons into the workforce. There were no major differences in mortality between workers whose jobs did or did not bring them into contact with radiation. In two studies, of workers at the Hanford plant in the USA and at the Sellafield plant in the UK, mortality from multiple myeloma was statistically significantly related to cumulative radiation exposure. In the study of employees of the United Kingdom Atomic Energy Authority, mortality from prostatic cancer was significantly related to cumulative radiation exposure, and also associated with being monitored for possible internal contamination by radionuclides. Despite the large number of workers studied, the data are still too sparse to permit precise statements to be made about the magnitude of the increase in cancer risk per unit dose associated with repeated exposure to low levels of ionizing radiation. The findings thus far are compatible with there being no increase in risk at all, or with an increase about ten times that predicted by the risk estimates of the International Commission on Radiological Protection (ICRP).

INTRODUCTION

It has long been recognized that, since some workers in the nuclear industry are repeatedly exposed to low levels of ionizing radiation, studying their health should make an important contribution to our understanding of its biological effects. Many epidemiological studies of workers are now underway both in the United Kingdom and abroad, identifying all who have been employed in a particular nuclear plant and following them to find out about their subsequent health and cause of death.

Almost ten years ago the first report of the results of the follow-up study of workers at the Hanford plant, in Washington State USA, was published by Mancuso, Stewart and Neale.[1] Their findings suggested that the actual risk of cancer associated with repeated exposure to low-level ionizing radiation was up to twenty times greater than that predicted from the experience of highly exposed populations such as the Japanese Atomic-bomb survivors.

STUDIES OF WORKERS IN THE NUCLEAR INDUSTRY

The United Kingdom Atomic Energy Authority Mortality Study

The United Kingdom Atomic Energy Authority (UKAEA) were concerned about the implications of the Hanford findings for their employees. They approached the Medical Research Council and asked for an independent group to be nominated to study the mortality of their workforce. The Epidemiological Monitoring Unit (EMU), at the London School of Hygiene and Tropical Medicine, was invited to carry out the study.

Figure 1 shows the location of the UKAEA establishments. Data collection for the study of its workforce began in 1980. In this, although the EMU's contract was with the Medical Research Council, we collaborated closely with staff at the UKAEA. Under our guidance they assembled information on the entire workforce employed between 1946 and 1979 including data on past radiation exposure.[2] The data collected were checked by the EMU, and assessed to be as complete and accurate as could reasonably be expected. Follow-up was carried out by the EMU, through the National Health Service Central Registers and the Department of Health and Social Security's Record Branch. All notifications of death and copies of death certificates were sent to the EMU directly and not passed onto the UKAEA until all the radiation records had been received by the EMU.

The analyses we reported in 1985 were based on approximately 40 000 employees, of whom all but 0.6 per cent had been followed up to the end of the study period, 31 December 1979 (Figure 2, Beral *et al.*[3]). The collective recorded whole-body radiation exposure was 660 Sieverts.

Studies of other workforces

A total of eight follow-up studies of workers in the nuclear industry have now been published, two of which are the studies of the Hanford and UKAEA

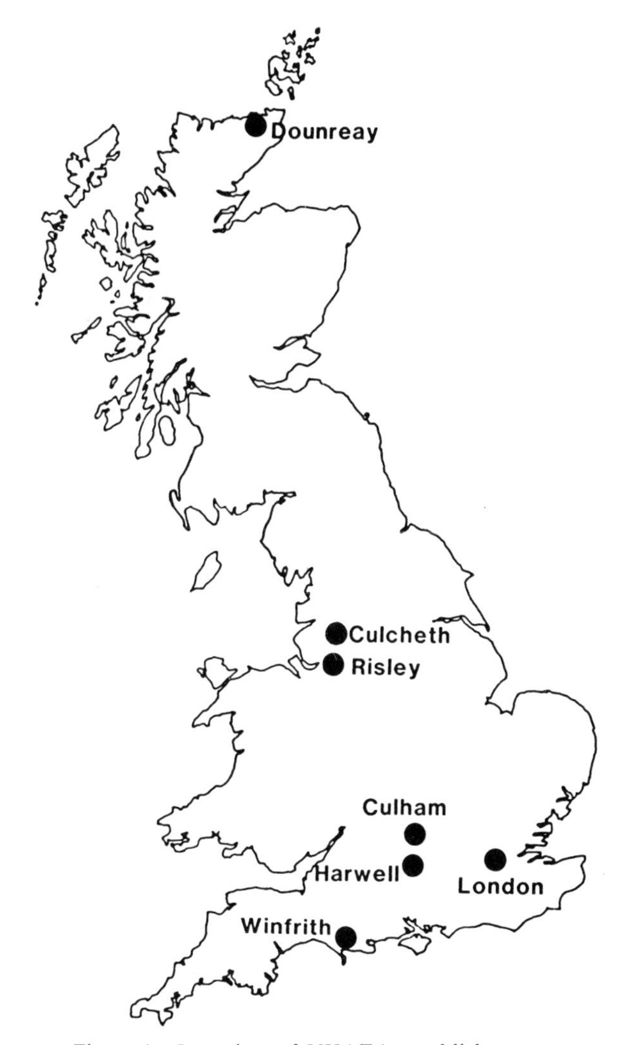

Figure 1 Location of UKAEA establishments.

employees. The eight studies are listed in Table 1 and together they comprise the experience of a total of 120 000 workers. All the studies have been of a similar design, and all have presented data on mortality of the workforce in comparison with death rates in the national or local community.

Even before the individual studies began, it was recognized that their findings alone were unlikely to permit firm conclusions about the precise magnitude of any increase in cancer risk associated with low-dose ionizing radiation. For example, in the whole population of 40 000 UKAEA workers,

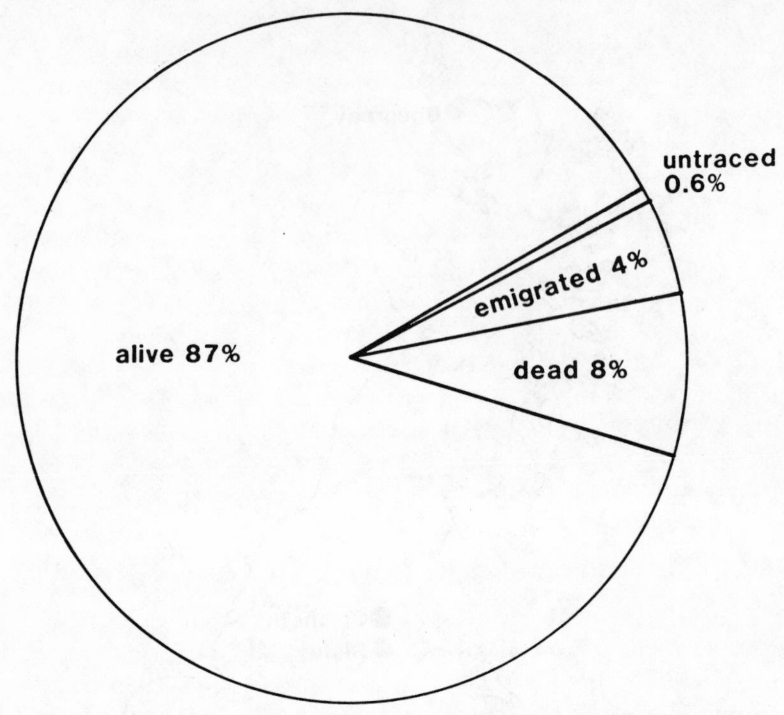

Figure 2 Summary of the status of 39 546 employees of the UKAEA who were followed up (to December 1979) to identify those who had died.

Table 1 Studies of workers in the nuclear industry

Nuclear establishment	Country	Size of workforce	Reference
Hanford Plant	USA	17 000	(7)
Portsmouth Naval Shipyard	USA	25 000	(8)
Rocky Flats Weapons Plant	USA	5 000	(6)
United Nuclear Corporation	USA	4 000	(9)
Oak Ridge National Laboratory	USA	8 000	(10)
Pantex Weapons Plant	USA	4 000	(11)
United Kingdom Atomic Energy Authority	UK	40 000	(3)
Sellafield Plant, British Nuclear Fuels	UK	14 000	(4)

only 28 deaths from leukaemia would be predicted on the basis of national mortality rates at ages 15–74 years. Because of chance variation in disease rates, any number between 19 and 40 deaths would be consistent with the baseline leukaemia rates in the community. In addition, on the basis of the recorded radiation exposure in the UKAEA workforce, only one extra death from leukaemia would result if the existing risk estimates of the ICRP (based largely on the backward extrapolation from highly exposed individuals) are correct. Moreover, one additional leukaemia death would be indistinguishable from the usual occurrence of leukaemia in the UKAEA workforce. Only if the ICRP figures were 20 times too low—so that more than 40 leukaemia deaths occurred—would the mortality be clearly greater than expectation. A similar conclusion could also be drawn for mortality from other malignancies. Furthermore, it was recognized before the studies began that the workforces were likely to be of higher social class and healthier than their respective national populations, and thus lower overall mortality rates than the national or local averages would be expected.

MORTALITY OF WORKERS IN THE NUCLEAR INDUSTRY

Table 2 summarizes the mortality from all causes in the eight workforces whose follow-up has been completed so far. Mortality is expressed as a standardized mortality ratio (SMR), which is the number of deaths observed in the workforce, expressed as a percentage of that expected on the basis of national mortality rates. It can be seen that in all studies the SMR is below 100, although there is some variation from one workforce to another. Overall the

Table 2 Standardized mortality ratios (SMRs) for certain causes of death in the eight studies of workers in the nuclear industry (see Table 1 for references)

Establishment	SMR (number of deaths)		
	All causes	All cancers	Leukaemia
Hanford Plant	80 (3994)	86 (733)	59 (19)
Portsmouth Naval Shipyard	89 (4762)	94 (977)	94 (39)
Rocky Flats Weapons Plant	64 (334)	75 (79)	117 (5)
United Nuclear Corporation	82 (216)	85 (51)	84 (2)
Oak Ridge National Laboratory	73 (966)	78 (194)	149 (16)
Pantex Weapons Plant	72 (269)	60 (44)	128 (4)
United Kingdom Atomic Energy Authority	74 (2856)	79 (827)	123 (35)
Sellafield Plant, British Nuclear Fuels	98 (2277)	95 (572)	63 (11)
Total	82 (15674)	86 (3477)	90 (131)

pooled SMR is 82, based on 15 674 deaths. Within the workforces, the SMRs were similar in radiation workers and in others who were not monitored because their work did not bring them in contact with radiation. In the UKAEA Mortality Study the SMR was 75 for those monitored for exposure to radiation and 74 for those not monitored;[3] in the British Nuclear Fuels (BNF) Mortality Study, the SMR was 95 in radiation workers and 107 in other workers.[4] Thus there was no suggestion that the radiation workers experienced an unduly high mortality.

Standardized mortality ratios for death from all cancers are also shown in Table 2. As for all causes of death, the eight SMRs are all below 100, although they vary from study to study. The overall pooled SMR is 86, based on 3477 cancer deaths. There were no major differences in mortality between radiation workers and other employees, either in the UKAEA workforce (SMRs of 76 and 83 respectively), or in the BNF workforce (SMRs of 95 and 95, respectively).

No cancer at any specific anatomical site was significantly increased at any single nuclear establishment, nor were there major differences in mortality from specific cancers in radiation workers and the remainder of the workforce. Table 2 also shows the SMRs for leukaemia in the eight workforces. There is more variability than in the results for all cancers, presumably because of the small number of deaths from leukaemia at any one plant. The overall pooled SMR is 90, based on 131 deaths.

MORTALITY IN RELATION TO EXTERNAL RADIATION EXPOSURE

Data on mortality according to cumulative whole-body exposure have been reported for the UKAEA, Sellafield and Hanford workforces.[3,4,5] In general there was no statistically significant trend of increasing mortality from all cancers with increasing exposure. For the UKAEA workforce, Figure 3 shows the ratio of observed to expected deaths from all cancers in relation to recorded whole-body exposure. The numbers of deaths observed are shown above each bar, and can be seen to be small, especially for the most highly exposed workers with cumulative whole-body exposures of 100 mSv or more. Although there is a suggestion that mortality increases with increasing exposure, the trend is far from being statistically significant. The estimated increase in mortality from all cancers, calculated from these data gave a value of 12.5 extra cancer deaths per million person-years per rem (10 mSv). This is about four to five times the ICRP figure, but the slope of the line through the points has sufficient variability to be consistent with no increase in risk at all, or with a risk fifteen times as great as the ICRP figures. The Sellafield and Hanford data are also based on small numbers so that the findings are compatible with no effect at all or with increases in cancer risk an order of magnitude greater than the ICRP figures.

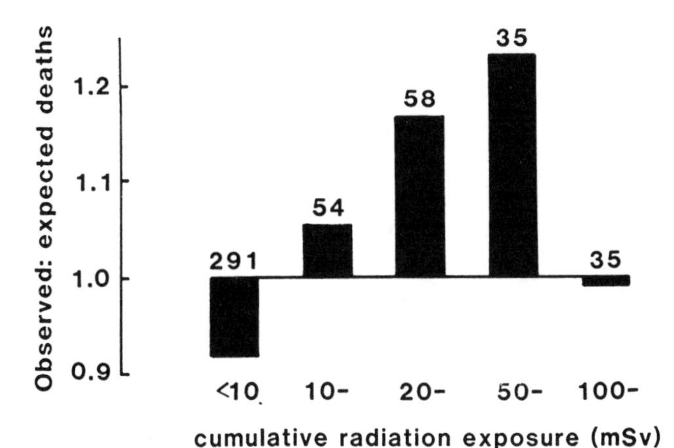

Figure 3 Relationship of mortality from all malignant neoplasms and final cumulative radiation exposure in the UKAEA workforce (1946–1979). (The numbers above the columns are the number of cancer deaths in each group.)

The same general conclusions also apply to mortality from leukaemia (Figure 4). The only condition significantly related to external radiation exposure in more than one nuclear establishment was multiple myeloma. In both the Sellafield and the Hanford workforces, the trend of increasing mortality with increasing exposure was statistically significant although there were few deaths—two and three respectively—in the most highly exposed groups

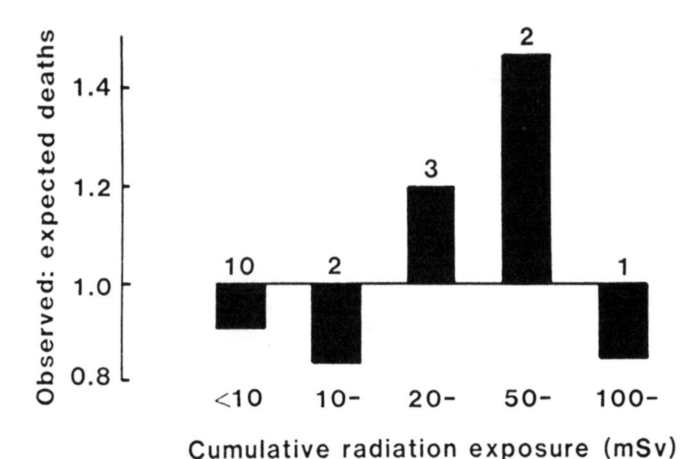

Figure 4 Relationship of mortality from leukaemia and final cumulative radiation exposure in the UKAEA workforce. (1946–1979). (The numbers above the columns are the number of deaths from leukaemia in each group.)

(Figure 5). In the UKAEA workforce the trend was in a similar direction but not statistically significant. The only statistically significant trend with increasing levels of radiation exposure found in the UKAEA workforce was for prostatic cancer. This was an unexpected finding since no similar trends were evident either in the Sellafield or the Hanford workforces. There are, however, certain observations which suggest that this may not be a chance finding. Although prostatic cancer mortality was high among the most heavily exposed workers, it was particularly high among the men monitored for possible internal contamination by radionuclides, especially tritium (Table 3). There may be a specific radionuclide, or other chemical, which is concentrated in the

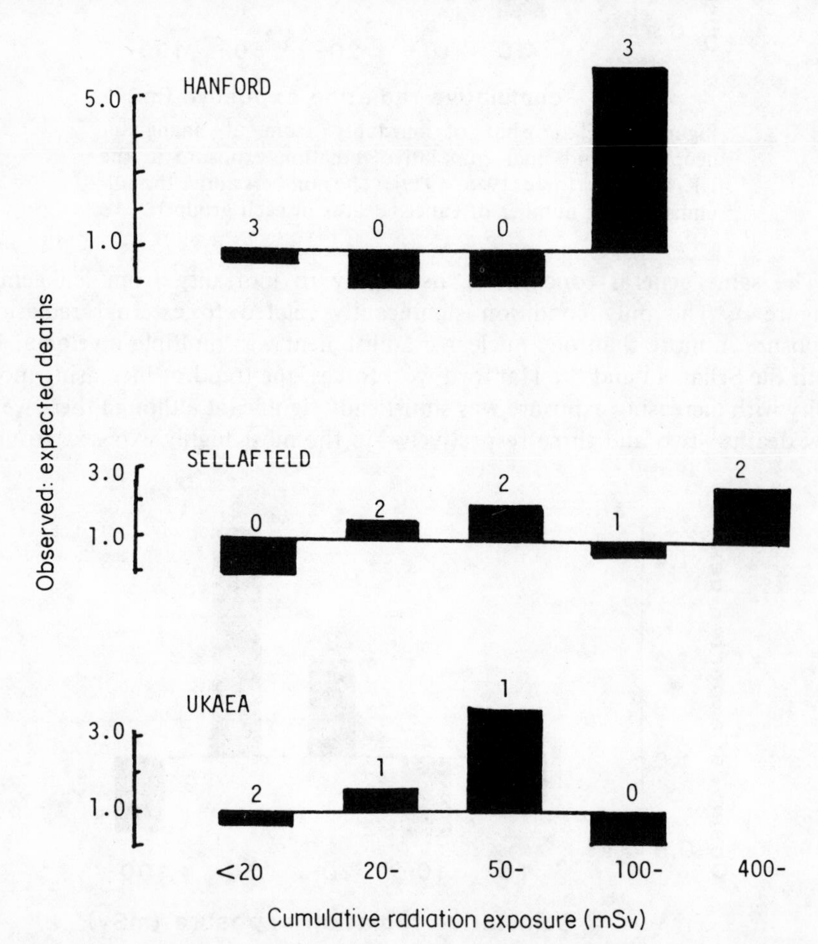

Figure 5 Relationship of mortality from multiple myeloma and radiation exposure in the Hanford, Sellafield and UKAEA workforces. (The numbers above the columns are the number of deaths from multiple myeloma in each group.)

Table 3 Standardized mortality ratios (SMRs) for prostatic cancer in United Kingdom Atomic Energy Authority radiation workers[2]

Type of exposure	SMR (number of deaths)
External cumulative whole body dose:	
less than 50 mSv	70 (10)
more than 50 mSv	385[a] (9)
Monitored for possible internal contamination by:	
tritium	889[a] (6)
plutonium	153 (2)
other (unspecified) radionuclides	254[a] (9)

[a] $p < 0.01$

prostate and induces malignancy in that organ. The carcinogen is unlikely to be tritium, because tritium is not concentrated in the prostate, but it may be some substance which is often used in conjunction with tritium. We are currently collaborating with the UKAEA in investigating this association further. The Sellafield study has not analysed separately groups monitored for internal contamination by radionuclides, so we cannot compare our findings with theirs. At Hanford, few workers are exposed to tritium, and none of the men who died from prostatic cancer had been monitored for tritium contamination (Gilbert, personal communication).

Approximately 3500 UKAEA employees had been monitored for plutonium exposure, and no unusual findings were noted in that group.[2] The only other study to have considered plutonium workers separately was that by Wilkinson *et al.*[6] They noted an increase in benign intracranial tumours, which was not evident, however, in the workers most heavily exposed to plutonium.

CONCLUSION

So far no major health hazard associated with working in the nuclear industry has been identified. If the radiation to which the workers are exposed is causing any increase in cancer, its effects are so small as to be indistinguishable from the normal fluctuations in disease occurrence. Multiple myeloma is the only condition where a consistently increased mortality has been noted in the most highly exposed workers—but in each workforce there were probably no more than two or three extra deaths from this condition in the entire workforces.

Despite these reassuring findings it is also true that the studies described do not permit any definitive statement to be made about the validity of the existing ICRP risk estimates, other than that they are unlikely to be out by a factor of 20. The findings so far are consistent with there being no increased cancer risk at all, and, at the same time, with a risk ten to fifteen times the ICRP figures. To lessen the uncertainty around the estimates of the increase in cancer risk per unit dose at low levels of ionizing radiation, it will be necessary to pool the results from several studies; this we are in the process of doing.

ACKNOWLEDGEMENTS

We thank Helen Edwards for typing the manuscript.

REFERENCES

1. T. F. Mancuso, A. Stewart, and G. Kneale, Radiation exposure of Hanford workers dying from cancer and other causes, *Health Phys.*, **33**, 369–84 (1977)
2. P. Fraser, M. Booth, V. Beral, H. Inskip, S. Firsht and S. Speak, Collection and validation of data in the United Kingdom Atomic Energy Authority mortality study, *Br. Med. J.*, **291**, 435–9 (1985).
3. V. Beral, H. Inskip, P. Fraser, M. Booth, D. Coleman and G. Rose, Mortality of employees of the United Kingdom Atomic Energy Authority, 1946–1979, *Br. Med. J.*, **291**, 440–47 (1985).
4. P. G. Smith and A. J. Douglas, Mortality of workers at the Sellafield plant of British Nuclear Fuels, *Br. Med. J.*, **293**, 845–54 (1986).
5. E. S. Gilbert, and G. R. Peterson, A note on 'Job related mortality risks of Hanford workers and their relation to cancer effects of measured doses of external radiation'. *Br. J. Ind. Med.*, **42**, 137–9 (1985).
6. G. S. Wilkinson, G. L. Volez, J. F. Acquavella, G. L. Tietjen, M. Reyes, R. Brackbill and L. D. Wiggs, Mortality among plutonium and other workers at a nuclear facility. *Los Alamos National Laboratory Document LA-UR-83-266*, 1983, Los Alamos, New Mexico.
7. E. S. Gilbert, and S. Marks, An analysis of the mortality of workers in a nuclear facility, *Radiate. Res.*, **79**, 122–48 (1979).
8. R. A. Rinsky, R. D. Zumwalde, R. J. Waxweiler, W. E. Murray Jr., P. J. Bierbaum, P. J. Landrigan, M. Terpilak and C. Cox, Cancer mortality at a naval nuclear shipyard, *Lancet* **1**, 231–5 (1981).
9. O. C. Hadjimichael, A. M. Ostfeld, D. A. D'Atri and R. E. Brubaker, Mortality and cancer incidence. Experience of employees in a nuclear fuels fabrication plant, *J. Occup. Med.*, **25**, 48–61 (1983).
10. H. Checkoway, R. M. Mathew, C. M. Shy, J. E. Watson Jr, W. G. Tankersley, S. H. Wolf, J. C. Smith and S. A. Fry, Radiation work experience and cause-specific mortality among workers at an energy research laboratory, *Br. J. Ind. Med.*, **42**, 525–33 (1985).
11. J. F. Acquavella, L. D. Wiggs, R. J. Waxweiler, D. G. MacDonell, G. L. Tietjen and G. S. Wilkinson, Mortality among workers at the Pantex weapons facility, *Health Physics*, **48**, 735–46 (1985).

Radiation and Health
Edited by R. Russell Jones and R. Southwood

9

Occupational Exposures and the Case for Reducing Dose Limits

DAVID GEE
National Health and Safety Officer,
General, Municipal, Boilermakers and Allied Trades Union

INTRODUCTION

The General, Municipal, Boilermakers and Allied Trades Union has members exposed to radiation in nuclear waste reprocessing, nuclear fuel fabrication, nuclear power stations, nuclear submarine construction and maintenance, hospital X-ray and nuclear medicine departments and factories using radiation sources. Other members are exposed via industrial radiography on construction sites, non-coal underground mines, waste disposal and even handling geological specimens in museums. Since 1981 we have criticized the narrow representation on the International Commission on Radiological Protection (ICRP), the self-appointed group of experts which recommends radiation exposure limits for countries to adopt, and have questioned the acceptability of their limits. In 1983[1] we submitted our case for a five-fold reduction in dose limits to the Sizewell Inquiry into PWR construction.

This chapter describes the GMB approach to all harmful agents encountered in either the workplace or the general environment; summarizes current radiation exposures in the UK and our arguments for a five-fold reduction in dose limits; and concludes with a summary of the only agreed compensation scheme in the world for radiation-induced cancer amongst workers.

APPROACHES TO THE QUESTION OF PROOF OF HARM

We start our approach to harmful agents by noting the bitter lessons learnt from the history of asbestos, mule spinning oil cancer, aromatic amines, etc., where the *traditional approach* of giving the benefit of any scientific doubt to the inanimate agents, rather than to people, has resulted in thousands of unnecessary deaths; and in high community costs from dead and disabled people, from harmful wastes, and from unplanned retrofit controls on industry. The *traditional approach* assumes that agents are harmless until proven harmful by predominantly human evidence, assembled in such significant quantities and configurations as to meet the high 'beyond all reasonable doubt' level of proof required. By the time the criteria of this approach have been met there are not only too many dead people but there is also a 'pipeline' of future victims who have been exposed before adequate protection occurred. The 20–40 year latent period for the chronic effects of these recognized harmful agents means that many future deaths in the 'pipeline' are inevitable even after the harmful agent has been banned or adequately controlled. The case of asbestos best illustrates the point, with a current annual death toll of 2000 people a year in the UK and with probably 50 000 more to come over the next 30 years[2] whether we ban asbestos or not.

An alternative approach to the question of proof of harm is clearly necessary and since 1980 we have been advocating the *trade union* approach which gives the benefit of the doubt to people, rather than to inanimate agents in the workplace. We start by assuming that agents are more likely to be harmful rather than harmless; then put more weight on evidence from structural similarity, from laboratories (e.g. mutagenicity tests, animal tests), and from workers experience, than on negative epidemiology;* and then weigh all the evidence against the lower 'balance of probabilities' level of proof, used in our civil courts, rather than the 'beyond all reasonable doubt' level of the criminal courts.

An indication of how these two approaches work out in practice is afforded by applying them to two contemporary issues, passive smoking and mineral wool insulation. If the traditional approach is applied neither of these two agents can be regarded as a carcinogen, but if the union approach is adopted then both are lung carcinogens.†

The application of the trade union approach to radiation justifies lowering the maximum permitted exposure limits by five-fold as the evidence below demonstrates.

*Richard Peto, in testimony to the US Congress illustrated the weakness of most negative epidemiology, i.e. human studies that do not show a health risk, by saying in the discussion on the limitations of non positive studies:

> If in theory you have observed 50,000 people of about 40 years and you can distinguish well between those who are exposed and those who are not exposed, then you can start making statements which should not turn out to be 'false negatives'. *Federal Register*, Vol. 45, No. 15, 22 Jan. 1980, p. 5050.

†For the evidence on passive smoking see the Oxford Scientific Consensus Conference (21/22 March 1986), Green College, Oxford. For mineral wool see the WHO Conference report on the Copenhagen Conference on MMMF October 1986.

The usual procedure for determining the exposure limits for any harmful agent is to consider the health effects, then the current exposure data, then combine the two in the essentially political process of determining the acceptability of the risks that flow from choosing particular exposure limits‡ that do not place unacceptable burdens on industry.

THE CASE FOR REDUCING EXPOSURE LIMITS

Our case rests on eleven different arguments most of which independently justify a two to five-fold reduction in exposure limits.

1. Uncertainty

An important conclusion from all scientific reviews of the risk from low-level radiation is that the risk estimates 'are based on incomplete data and involve a large degree of uncertainty, especially in the low dose region'.[3] Therefore one should err on the side of caution when choosing exposure limits.

2. Risk estimates embrace a range of possible answers

All major scientific reviews, except that of ICRP, give a range of answers in their risk estimates (Table 1).

Uncertainty demands a range of estimates: prudence and caution demands choosing the upper end from the range of possible risks, if people are to be given the benefit of the scientific doubt.

3. Experts question ICRP risk estimates

In comparison with the period from 1930–75 which was largely one of consensus on radiation risks, experts have questioned ICRP risk estimates. This has occurred because uncertainty has increased, following the emergence of both new data and new interpretations, and because many did not agree with the new procedures and recommendations of ICRP 26 (1977), which are the basis

‡For example, the risk from asbestos at the current UK exposure limit of 500 000 fibres/m^3 (0.5 f/ml) are 2–3 cancer deaths/100 exposed to a lifetime (40 years) at the limit. This risk is unacceptably high, but was necessary to avoid the closure of the 'dirty end' of asbestos production, textiles. The GMB does not accept the validity of allowing the 'dirty end' of an exposure group to determine the exposure limit because it imposes unnecessary high risks on the majority of users who can get well below the limit. Although they are supposed to get 'as low as is reasonably practicable' (ALARP) below the limit, this, like the similar concept in radiological protection (ALARA), is largely unenforceable. We propose giving time limits for the 'dirty end' to reach the lower exposure limit that others can reach immediately. See GMB Policy Paper 'A Trade Union Approach To Toxic Substances'.

Table 1

	Cancer deaths (per million persons per rem per year)
BEIR I(1972)	115–621
UNSCEAR (1977)	75–175
ICRP 26 (1977)	125
BEIR III (1980)	158–501

of ICRP's current exposure limits. For example, Professor Radford, ex-chair of BEIR III recommends a ten-fold reduction,[4] largely on the basis of the latest Japanese Bomb data; Stewart Mancuso and Kneale[5, 6] recommend a 10–15-fold reduction, largely on the basis of the data from the US reprocessing plant at Hanford (USA); Professor Morgan,[7] ex-ICRP, recommends a two-fold reduction; Professor Rotblat[8] recommends a five-fold reduction, largely on the basis of re-interpreting the Japanese Bomb data; and several other experts consider that ICRP have seriously under-estimated the risks. We agree with most of what those experts say and consider a five-fold reduction to be a reasonable compromise between these different opinions, which in the studies cited above, do not take into account the Japanese dosimetry revision described below.

4. Cancer induction is likely to follow the relative risk, not absolute risk model

ICRP chose to use the Absolute Risk (AR) model in 1977 but since then the latest data from Japan and elsewhere suggests that the Relative Risk (RR) model is more likely to represent the risk of cancer induction. As the RR model rises proportionately with the rising background risk of cancer, the overall lifetime risk is greater than that predicted from the AR model, where risk is a fixed constant, irrespective of the background risk. The lifetime risk difference, from 20 years of age, is estimated at 1.5–2.0 times greater for the RR model.

5. New data from the Japanese bomb data

The revision to the estimated radiation doses received by the Japanese people in Hiroshima and Nagasaki, which has emerged since the publication of ICRP 26, looks like producing an increase of 1.5–2.0 in the risk estimates.

In addition, the latest cancer statistics from the Bomb victims, especially the morbidity figures, are suggesting greater risks than ICRP estimates, as Professor Radford illustrates in Chapter 7 (this volume).

6. Non-bomb data may yield higher risks

Ever since Dr Alice Stewart's Oxford Survey,[9] which showed cancer in the children of mothers who had been X-rayed during pregnancy, there have been those who have questioned the validity of using the Japanese Bomb data as the main evidence for estimating the effects of long-term, low-level radiation in workers. One criticism points to the late assembly of the Japanese study population (1950) and suggests that the absence from that group of those who had survived the initial blast but who succumbed to the disease and trauma of the first few years after the Bomb makes the study population unrepresentative.[10] Others have criticized the control groups, which also received some radiation.[11] The risk estimates from the non-bomb data reviewed by Lindop and Charles[12] in 1981 concluded with the range 100–440 deaths/million/rem compared with the ICRP estimate of 125.

Since that review the first large-scale studies of UK workers have been published, but the numbers involved are still too small to give reliable results. However, if the AEA study[13] were four times larger and the results similar, then ICRP under-estimates risk by up to 10 times. The BNFL Study[14] is even smaller, allowing conclusions from the range that there is either no radiation risk from low level radiation or that ICRP under-estimates risk by up to 10 times.

7. Risk estimates should include cancer morbidity, not just cancer mortality

The cancer risk estimates built into the radiation dose limits are based on *cancer deaths*. Many cases of radiation-induced cancer do not always kill, especially skin, breast and thyroid. Getting cancer, however, is not trivial, causing obvious personal and social harm. The dose limits should therefore reflect the risk of *getting cancer* ('incidence') rather than just the risk of *dying from cancer*, which is lower. The CEGB/Lindop paper cited above shows that the non-bomb evidence supports cancer incidence risk rates that are *five times the ICRP figure*. For example, the ICRP risk estimate for death from thyroid cancer is one-twentieth of the risk of getting a thyroid cancer that will not be fatal. The union does not accept that getting thyroid, or any treatable cancer, should be excluded from the risk estimates of radiation work.

8. Exposure limits should protect the minority, not the 'average'

The official limits are based on the 'average' person and no account is taken of the biological differences between workers, except for teratogenic effects, even though there is considerable evidence about these differences. For example:

(i) Experimental animals show marked differences in response to radiation between species, and even between strains within species, and human biological variability is much greater than amongst experimental animals;

(ii) Women have a greater risk of cancer from radiation than men, in addition to the added risk of teratogenic effects on the fetus. Professor Radford puts this cancer risk difference between the sexes at about three times;

(iii) Young workers are also at greater risk than older workers;

(iv) Five to ten per cent of the population have deficiencies in DNA repair mechanisms which make them especially sensitive to radiation damage.

On general illustration of this biological variability is the case of cancer in one member who received compensation on the evidence submitted, but who did not receive the maximum permissible does for the 'average' worker.

If we are to protect those susceptible sub-groups within the human population, which undoubtedly exist, the dose limits should be lowered from the level suitable for the 'average' worker to a fraction of that. The limits for radiation effects such as cataracts and skin damage are set to protect the most sensitive sub-groups, so why not the limits for the cancer effects? The control limits (TLV's etc.) for other harmful agents are set in this way. An arbitrary fraction of say one-half would at least give some protection to susceptible sub-groups, especially those receiving above average background or medical doses, which are not included in risk estimates.

9. ICRP limits impose unacceptable risks

The 2–3 percent excess lifetime risk from exposure to the maximum whole-body dose of 5 rem (50 mSv) is too high. Although only a small group of workers approach this annual dose (see below) this reinforces not weakens, the cased for reducing the limit towards more acceptable levels, i.e. nearer a 1 in 10 000 lifetime risk.

10. Lower limits reduce exposures more effectively than 'ALARA' does

Although employers are under a duty to reduce exposures as far below the maximum permitted limits as is reasonably achievable ('ALARA'), this is very difficult to achieve, or enforce, in practice, without other numbers for the engineers and health physics people to aim for, such as the one-tenth or three tenths of the limit used in radiological protection. Our experience of radiation, and other harmful agents which are governed by the duty 'as low as is reasonably practicable', is that lower limits are the most effective way of achieving reductions in exposure.

11. Cancer victims need just compensation

The pioneers of the nuclear industry, and of non-coal underground mining, who are today's cancer victims of past, higher, radiation exposures should have their chances of radiation-induced cancer assessed against the more likely dose/response data supporting lower exposure limits, rather than against the more conservative ICRP risk estimates. Current assessment procedures can be dominated by experts who argue that doses below the ICRP maximum permitted limits do not confer a compensatable cancer risk. Judges and others are inclined to accept their arguments, even though the limits for radiological protection do not take account of the risk to particular individuals, who can be towards the end of a biological spectrum of response to exposure. For example, we have had cases of lung cancer from radon in underground mines turned down because they have not exceeded maximum permitted doses, despite the latest evidence suggesting very high risks from the current dose limits.[15]

Taken together the above arguments more than justify a five-fold reduction in dose limits. Let us now look at current exposures to see whether such a reduction would cause any great difficulties for UK industry.

Current radiation exposures of UK workers

There are about 300 000 workers in the UK exposed to radiation at work, but only about 139 000 are routinely monitored for exposures. As Table 2 shows, their average annual exposure is comparatively low, with those in the nuclear fuel cycle averaging 2.3 mSv (230 m rem). However, some occupational groups receive significantly higher doses, particularly in nuclear fuel reprocessing and non-coal-mines (7 and 8.9 mSv average annual dose respectively). The percentage of workers above the new 15 mSv (1.5 rem) 'investigation level' in the Ionizing Radiation Regs 1986 is very low, as Table 2 indicates, although the percentage of collective dose above 15 mSv (1.5 rem) is a fairer expression of the effort required to reduce exposures to below the 10 mSv limit we are aiming for.

It is clear from Table 2 that a five-fold reduction in the maximum permitted dose limits from 50 mSv (5 rem) to 10 mSv (1 rem) would cause few problems for most UK employers, except those in nuclear fuel reprocessing and non-coal-mines, who could be given a longer time period to achieve the new limit.

The new limits would have to be achieved by engineering controls and improved work methods, rather than by spreading the existing dose amongst more workers. Our other dose limit recommendations, e.g. a lower limit for younger workers less than 35 years old; acumulative lifetime dose limit for workers of 450 mSv (45 rem); and a collective dose limit for each facility would assist in keeping exposures to more acceptable levels.

Table 2 Occupational radiation exposures, 1984/85

Type of work	Total no. of workers	Average annual dose (mSv)	% of workers above 15 mSv (1.5 rem) individual dose
Nuclear fuel cycle			
Fuel fabrication	2 819	1.3	<1
Fuel enrichment	1 175[a]	0.4	<1
Fuel reprocessing	5 362	7	12
Power stations	20 943	1.4	<1
Research	9 699	2.8	4
Ministry of Defence	15 000 (rounded)	2.4	No information available
Industry			
Radiography	7 000	1.7	1 (10 cases above 50 mSv)
Tritium workers	328	3.4	6
Amersham International	1 260	3.8	7
Other industrial	19 000	0.4	<1 (7 cases above 50 mSv)
Medical			
General	39 000	0.7	<1
Mining			
Coal	115 200	1.2	0
Non-coal	1 630	8.9	35
Total			
(For routinely monitored)	139 000	1.1	1
All occupationally exposed	296 000	1.2	—

[a] Includes Risley.
Source: NRPB R173 1984, plus some 1985 additions from S. Hughes, NRPB, Dec. 1986.

Compensation for radiation-induced cancer

Unlike all other UK workers, radiation exposed workers in the nuclear industry do not have to prove negligence in order to claim compensation for radiation-induced cancer from their employers. A legal duty of strict liability means that they only need to show, on the balance of probabilities, that their cancer was caused by radiation. Even so, the GMB's first few cases against British Nuclear Fuels Ltd took about six years to settle. In order to speed up and improve the efficiency of the process the GMB put forward the idea of a scheme that would cover all workers who had even worked for BNFL and who had died of any cancer, and which would use tables or graphs to plot the probability of each worker's tumour being radiation-induced, depending on

Table 3

Chance that radiation caused cancer	Amount of compensation	
10 percent or more	25 per cent	of full compensation at law
30 percent or more	50 per cent	
40 percent or more	75 per cent	
50 percent or more	100 per cent	

the tumour, the sex of the worker, the age, and the total radiation dose from employment. The Compensation Agreement was signed in 1982 and had by the end of 1986 processed 133 claims, providing compensation for eleven workers. The risk factors in the scheme are the result of negotiations based around ICRP 26. A particular feature of the scheme is the payment of a proportion of full compensation where the probability of radiation-induced is less than 50 per cent, as Table 3 above illustrates.

Seven of the eleven settlements so far have been at 50 percent or less. The scheme was also applied retrospectively to all workers who had ever worked for BNFL (or employers at the time, such as the Atomic Energy Authority) since the Windscale (now called Sellafield) site began in 1949. The scheme is now being renegotiated around new union proposals which extend the scheme to cancer morbidity, and which use age adjusted risk factors. Adjustments to an estimated exposure may also be necessary following recent work on the AEA employees.[16]

CONCLUSION

The risks from low-level radiation are now much more in the public eye, following Chernobyl, and the determination of risks by unrepresentative expert committees like ICRP is no longer acceptable. The ICRP must allow involvement of experts representing the most at risk in their deliberations on dose limits. Even if this is not accepted we believe that our case for an immediate five-fold reduction is strong, and urge ICRP to accept it.

Whether our case is reasonable or not depends on one's point of view, but we commend the views of George Bernard Shaw on what is reasonable behaviour:

> The reasonable man adapts himself to the world; the unreasonable one persists in trying to adapt the world to himself. Therefore all progress depends on the unreasonable man. (*Man and Superman*).

REFERENCES

1. *GMB Evidence To the Sizewell Inquiry* 1983. GMB Publications, Thorne House, Ruxley Ridge, Claygate, Esher, Surrey KT10 OTL.
2. Richard Peto, 'Letters', *New Statesman* 10 September 1982.
3. 'Biological Effects of Ionizing Radiation Report' (BEIR III), 1980. US Academy of Sciences.
4. E. Radford, Human health effects of low doses of ionising radiation: the BEIR III controversy, *Rad. Res.*, **84** 369–94, 1980.
5. T. F. Mancuso, A. L. Stewart and G. W. Kneale, Radiation of Hanford workers dying from cancer and other causes, *Hlth Phys.*, **33**, 369–85 (1977).
6. G. W. Kneale, T. F. Mancuso and A. M. Stewart, Identification of occupational mortality risks for Hanford workers, *Br. J. Ind. Med.*, **41**, 6–8 (1984).
7. K. Z. Morgan, Cancer and low level ionising radiation, *Bulletin of Atomic Scientists*, **34**, 30–40 (1978).
8. J. Rotblat, The risks for radiation workers, *Bulletin of Atomic Scientists*, **34**, 41–6 (1978).
9. A. M. Stewart, J. Webb and D. Hewitt, Preliminary communication: malignant disease in childhood and diagnostic irradiation in utero. *Lancet*, **2**, 497–8 (1956).
10. A. Stewart, Detection of late effects of ionising radiation: why deaths of A-bomb survivors are so misleading. *Int. J. Epidemiol.*, **14**, 52–6 (1985).
11. Inge Schmitz in Feurhake and Carbonell, 'Evaluation of low level effects in the Japanese A-bomb survivors after current dose revisions and estimation of fall out contributions, IAEA Symposium on Biological Effects of Low Level Radiation, April 1983.
12. Lindop and Charles, 'Risk assessment without the bombs, *J. Soc. Radiol. Protect.*, **1**(3), 1981.
13. V. Beral, H. Inskip, P. Fraser, M. Booth, D. Coleman and G. Rosie, Mortality of employees of the UKAEA, 1946–1979. *Br. Med. J.*, **291**, 440–7 (1985).
14. P. G. Smith and A. J. Douglas, Mortality of workers at the Sellafield plant of British Nuclear Fuels, *Br. Med. J.*, **489**, 845–54 (1986).
15. D. C. Thomas and K. G. McNeill, Risk Estimates for the Health Effects of Alpha Radiation (1982). Atomic Energy Control Board, PO Box 1046, Ottawa, Canada KIP 5SP.
16. H. Inskip, V. Beral, P. Fraser, M. Booth, D. Coleman and A. Brown, Further assessment of the effects of occupational radiation exposure in the UKAEA Mortality Study, *Brit. J. Ind. med.*, **44**, 149–60 (1987).

Radiation and Health
Edited by R. Russell Jones and R. Southwood

10

The International Commission on Radiological Protection—A Historical Perspective

ROGER J. BERRY
International Commission on Radiological Protection, Department of Oncology, Middlesex Hospital Medical School, London, UK

ABSTRACT

The International Commission on Radiological Protection (ICRP) was established in 1928, under the name 'International X-ray and Radium Protection Committee' by the Second International Congress of Radiology meeting in Stockholm. It assumed its present name and organizational form in 1950 in order to cover more efficiently the expanding field of radiation protection. The policy adopted by the Commission in preparing its recommendations is to consider the fundamental principles and quantitative bases upon which appropriate radiation protection measures can be based, while leaving to the various national protection bodies the responsibility of formulating the specific advice, codes of practice or regulations that are best suited to the needs of their individual countries. ICRP is composed of independent members, chosen on the basis of their recognized activity in the field of medical radiology, radiation protection, physics, health physics, biology, genetics, biochemistry and biophysics, with regard to an appropriate balance of expertise rather than nationality. The Commission and its Committees are fully independent of both national governments and the nuclear industry although ICRP has maintained its close relationship with medical radiology. With its Publication 26 (1977), the Commission embarked on a new system of radiological protection in which three elements, *Justification*, *optimization* and *dose limits* interact to keep doses to both occupationally-exposed persons and to the general public *As low as reasonably achievable, economic and social considerations being taken into account*.

Although Roentgen's historic discovery of x-rays took place in November 1895, the first published report in English appeared in the *Standard* newspaper of 7 January 1896 on which day the engineer, Campbell-Swinton produced the first radiograph taken in England. Anecdotal reports of sore eyes and redness of the skin in the early x-ray pioneers led to the first published report of x-ray damage to man in the *British Medical Journal* of 18 April 1896.[1] By 1900, the first of what were to be over sixty British martyrs had died from cumulative radiation exposure. My own hospital, the Middlesex, lost its first radiographer, Reginald Mann (in 1916) and its first radiologist, Cecil Lyster (in 1920) by death from radiation damage. However, it was the ninth British martyr, Ironside Bruce, radiologist at Charing Cross Hospital, whose death in 1921 at the early age of 36 led to public outcry including letters to *The Times* calling for immediate cessation of the use of x-rays because of their hazard to health. The Establishment responded by setting up the British X-ray and Radium Protection Committee, chaired by the then President of the Royal College of Physicians, Sir Humphry Rolleston. Its recommendations, presented to the First International Congress of Radiology in London in 1925, were adopted by the Second Congress in Stockholm in 1928 and included the setting up of an International Protection Committee. Its first recommendations stated that

> The dangers of over exposure to x-rays can be prevented by the provision of adequate protection and suitable working conditions. It is the duty of those in charge of x-ray and radium departments to ensure such conditions for their personnel. The known effects to be guarded against are: a) Injuries to the superficial tissues; b) Derangements of internal organs and changes in the blood.

dose-limitation was achieved by limits to working hours

> a) not more than seven working hours a day. b) not more than five working days a week. The off-days to be spent as much as possible out of doors. c) not less than one month's holiday a year.

The recommendations were firmly based in common sense

> An x-ray operator should on no account exposure himself to a direct beam of x-rays... .An operator should place himself as remote as practicable from the x-ray tube... . The x-ray tube should be surrounded as completely as possible with protective material of adequate lead equivalent.[2]

Appreciation of the the importance of radiation protection came only slowly, however. X-rays were used in medicine with little awareness of potential late consequences; children with fungal infections of the scalp such as ringworm were epilated with x-rays as part of their treatment, and a multiplicity of benign skin conditions were treated by superficial irradiation. With much more pressing clinical indications, ankylosing spondylitis (an excruciatingly painful disease and the second commonest cause, after trauma, of low back pain in young men) was effectively treated with deep x-rays. There was no legal control of who could use x-ray machines in most countries, however, and non-medical uses such as that of fluoroscopy in shoe-fitting and actual therapeutic use of x-rays in removing facial hair in 'beauty clinics' were by no means rare. Radium-containing waters were inhaled or drunk as part of 'spa' treatment—and in some countries they still are! Public awareness of the risks of radiation exposure largely followed the detonation of nuclear weapons at Hiroshima and Nagasaki in 1945, and the early demonstration of an increased rate of leukaemia in the bomb survivors led to epidemiological studies of the pioneer radiologists as 'occupationally-exposed persons'. March in 1950, reported a retrospective survey of US doctors' obituaries in the *Journal of the American Medical Association* and showed a ninefold excess of deaths from leukaemia among radiologists compared with other doctors[3] (Table 1). Later epidemiological studies, using more conventional methods, showed clearly that among British radiological pioneers *no cause of death other than malignant disease* appeared in excess, and that after 1921 the cumulative cancer risk decreased in successive cohorts as radiological protection practices improved[4] (Table 2). Even amongst US radiologists, in whom there appeared to be an excessive death rate from non-cancer causes amongst early cohorts, once effective chemotherapy for tuberculosis became available their only increased cause of death was from later incidence of malignant disease. By the 1950s cohorts even this was indistinguishable from cancer rates in other doctors[5] (Table 3). Further information on human cancer risks following radiation exposures came from epidemiological studies of patients such as ankylosing spondylitics[6] and women who underwent multiple fluoroscopy to monitor artificial pneumothorax used for the treatment of tuberculosis in the pre-antibiotic era[7]. Thus, quite independently from risk estimates arising from studies of the Hiroshima and Nagasaki bomb survivors, sizeable bodies of data were being accumulated about the long-term risks to human beings of prior exposure to ionizing radiations—and the only significant long-term risk of relatively low-level irradiation appeared to be a dose-related increase in the lifetime risk of developing malignant disease.

Table 1 Leukaemia in US radiologists (March[3]) 1929–48 inclusive

200 radiologists dead
14 leukaemia (4.68 per cent)
65 922 non-radiologist physicians dead
344 leukaemia (0.51 per cent)
Nine fold excess leukaemia deaths among radiologists

Table 2 Cause of deaths of 339 pioneer radiologists (Smith and Doll[4])

Cause of death	Observed	Expected	O/E
Pre 1921			
All causes	339	324	1.05
All neoplasms	64	45	1.4
Leukaemia	4	0.65	6.2
Ca skin	6	0.77	7.8
Ca lung	8	3.7	2.2
Ca pancreas	6	1.9	3.2
All other causes	275	279	1.0

Cause of deaths of 999 pioneer radiologists who joined the British Institute of Radiology from 1921–54 (Smith and Doll[4])

Cause of death	Observed	Expected	O/E
1921–35			
All neoplasms	48	52	0.93
Leukaemia	3	1.3	2.3
Ca skin	2	0.6	3.5
Ca lung	13	12	1.1
1936–54			
All neoplasms	24	39	0.61
Leukaemia	1	1.3	0.77
Ca skin	0	0.4	—
Ca lung	10	12.2	0.82

In 1950, the former International X-ray and Radium Protection Committee was reorganized and renamed the International Commission on Radiological Protection. Although it retained its close links to the International Congress of Radiology, along with the International Commissions on Radiation Units and Measurements (ICRU) and Radiological Education (ICRE), it exercised greater autonomy to reflect the growing importance of radiation protection *not* associated with medical uses of ionizing radiations. It now consisted of a Main Commission, with a chairman and up to twelve other members, of whom a prescribed number retired before each four-yearly International Congress of Radiology. The work of the Commission was aided by four standing committees; one of these (Committee 1) was given the continuing remit to review all published evidence of radiation risk and to advise the Main Commission if its current recommendations failed to give an adequate margin of safety. Another standing committee (Committee 4) was asked to review how the Commissions' recommendations were incorporated into national regulations. Some 75 scien-

Table 3 Age and time-adjusted death rate among US radiologists (Matanoski *et al*[5])

Cohort	All causes	Death rate/10^3 person-years	
		Cancer	All except cancer
1920–29 (all ages):			
Radiologists	19.9	3.2	16.7
Physicians	16.9	2.1	14.8
ENT/ophthalmologists	13.2	1.7	11.5
1930–39 (to age 74):			
Radiologists	12.8	2.5	10.3
Physicians	11.7	1.6	10.1
Ophthalmologists	8.9	1.4	7.5
1940–49 (to age 64):			
Radiologists	6.5	1.6	4.9
Physicians	6.3	1.2	5.1
Ophthalmologists	5.6	0.5	5.1

tists and doctors from twenty countries now comprise the ICRP 'family' of Main Commission and Committees.

During the 1950s and 1960s, the Commission's recommendations on Maximum Permissible Annual Doses were translated into working rules in many countries. However, with increasing information from epidemiological studies about long-term radiation risks to man, there was a steady trend towards reducing their numerical values; the situation in the United Kingdom is shown in Table 4. Subsidiary limits were introduced to reduce lifetime occupational radiation dose accumulation, but by and large remaining below the weekly, quarterly or annual maximum permissible dose was used operationally as an index of having achieved a satisfactory standard of radiation protection; there was little pressure to reduce doses to occupationally-exposed personnel. Therefore, in 1977 the ICRP introduced in its Publication 26[8] a new system

Table 4 Dose limits (whole body) occupationally-exposed persons

1951–54	0.5 R/week
1955–59	0.3 R/week (200 R in a lifetime, averaging 5 R/year)
1959–77	5(N-18) R (rem)/year (3 R (rem)/13 weeks)
1977–	50 mSv (5 rem)/year)

of radiation protection based on three interrelated elements; *justification* of the practice which requires the use of radiation, *optimization* of the radiation dose so that it is 'as low as reasonably achievable, social and economic considerations being taken into account, and *dose limits* which now represent the lower bound of a region of totally unacceptable practice. The system was designed to avoid completely the *non-stochastic* effects of radiation—those effects such as the skin damage seen in the early workers and radiation-induced cataract—while keeping harmful *stochastic* effects (cancer) at an acceptably low level of risk. Weighting factors were introduced to ensure that the risk of stochastic effects would be equal whether the whole body was irradiated uniformly or non-uniformly; these factors differed fundamentally in concept from the maximum permissible doses to individual organs in earlier recommendations. In preparing its new recommendations, the Commission chose to use human data on radiation-induced leukaemia as its primary estimate of risk and to calculate the total expected risk of fatal cancer at about five times the number of leukaemias; this ratio was based primarily on the distribution of excess cancers in the early radiologists. The Commission also decided that for sparsely-ionizing radiations such as x- or gamma-rays the necessary interpolation between effects observed at high doses and those predicted at low doses (there can never be data for zero dose because all humans are exposed to natural background radiation including the radioactivity of the elements of their own bodies) should have some allowance for non-linearity of dose–response, with the vast majority of biological evidence to date suggesting that the dose–response would be concave upwards. The Commission therefore adopted an estimate of leukaemia risk per unit dose based on results at the lower end of the observed ranges and selected a value of 20×10^{-4} Sv^{-1}. Using the ratio of approximately five fatal cancers for each observed leukaemia, *vide supra*, this led to a final 'best' estimate for risk of all fatal malignancies of 125×10^{-4} Sv^{-1}. While this method of estimating risk is fairly robust against changes in models used to predict numbers of cancers yet to appear in exposed populations whose members have not all died, it is sensitive to changes in risks of radiation-induced leukaemia and re-evaluation of the dosimetry on which the Hiroshima and Nagasaki risk estimates were made might well increase this estimate by as much as a factor 1.5. The same dosimetric re-evaluation has made it clear that we have no estimates of late risk of human exposure to densely-ionizing radiations which are of the same quality as our risk estimates for x- and gamma-radiation. Recognizing these two problems, the Commission at its meeting in Paris in 1985 recommended that as interim measures a change should be made in the quality factor used for calculating dose equivalent for fast neutron irradiation, and that for members of the general public the principal dose limit should as a matter of prudence be reduced to 1 mSv, although in any year the prior limit of 5 mSv could be used as long as the cumulated lifetime dose does not exceed

70 mSv—and subject as always to the injunction that all doses should be kept *as low as reasonably achievable.*[9]

In 1987 the current ICRP recommendations will be ten years old, and as you would expect the Commission is reviewing whether they need modifying. At the meeting with its committees in Como in September 1987, the Commission will decide whether the overall system of dose-limitation itself needs to be revamped. We have to survey not only how new data on human beings and on other biological systems affect our quantitative recommendations, but also to consider the extent to which the Commission's recommendations have been incorporated successfully into national rules. I believe that the current ICRP system of dose limitation DOES encourage dose reduction to the minimum which is socially and economically justifiable. The lineal descendance of ICRP form a body founded to protect workers—radiologists—against radiation injury means that it has maintained its independence and its scientific integrity in formulating its recommendations. I believe that its recommendations are neither absurdly restrictive nor irresponsibly relaxed. I hope that now and in the future the Commission will retain your confidence in its impartiality and level-headed common sense—confidence which I believe it well deserves.

REFERENCES

1. S. Rowland, Report on the application of the new photography to medicine and surgery. *British Medical Journal*, **i**, 99–100 (1896).
2. International Recommendations for X-ray and Radium Protection, *British Journal of Radiology* (*new series*) **1**, 358–63 (1928).
3. H. C. March, Leukemia in radiologists in a 20-year period. *American Journal of Medical Science*, **220**, 282–6 (1950).
4. P. G. Smith and R. Doll, Mortality from cancer and all causes among British radiologists. *British Journal of Radiology*, **54**, 187–94 (1981).
5. G. M. Matanoski, R. Seltser, P. E. Sartwell, E. L. Diamond and E. A. Elliott, The current mortality rates of radiologists and other physician specialists: deaths from all causes and from cancer. *American Journal of Epidemiology*, **101**, 188–98 (1975).
6. P. G. Smith and R. Doll, Mortality among patients with ankylosing spondylitis after a single treatment course with X-rays. *British Medical Journal*, **284**, 449–60 (1982).
7. J. D. Boice and R. R. Monson, Breast cancer in women following repeated fluoroscopic examinations of the chest. *Journal of the National Cancer Institute*, **59**, 823–32 (1977).
8. International Commission on Radiological Protection, Publication 26, Recommendations of the International Commission on Radiological Protection. *Annals of the ICRP* **1**, 3, Oxford, Pergamon Press, 1977.
9. International Commission on Radiological Protection, Statement from the 1985 Paris meeting of the ICRP, *Annals of the ICRP.*, **15**, 3, Oxford, Pergamon Press, 1985.

Radiation and Health
Edited by R. Russell Jones and R. Southwood

11

ICRP Risk Estimates—An Alternative View

KARL Z. MORGAN
Former Director,
Health Physics Division,
Oak Ridge National Laboratory, USA

ABSTRACT

For 58 years ICRP has served as the international source of information on risks of exposure to ionizing radiation and has provided recommendations for radiation protection. In general its publications have served a very useful purpose of reducing unnecessary radiation exposure but in some respects ICRP has delayed action to reduce excessive exposure, has underestimated radiation risks and has recommended radiation exposure levels that are much too high. For decades it showed concern to reduce exposure of doctors and nurses but ignored the principal source of population exposure, namely, patient exposure. Beginning in 1960 we became aware of two serious radiation exposure problems (occupational exposure in uranium mines and population exposure from testing of nuclear weapons). One might have expected ICRP to be the first to try to reduce these exposures but it was conspicuous by its silence. In 1958 ICRP set limits of exposure for radiation workers and members of the public. Nineteen years later (1977) when it was realized that the risk of radiation induced cancer was ten to thirty times what it was perceived to be in 1958, ICRP might have been expected to recommend a major reduction in permissible exposure levels, but to the dismay of some of us, it increased them. It was also a great disappointment when in 1977, levels of MPC or radionuclides in air, water and food were increased for a large fraction of the more dangerous radionuclides. The reactor accident at Chernobyl calls for a number of new ICRP recommendations. When can we expect them?

The International Commission on Radiological Protection, ICRP, has been in existence for almost 60 years, beginning under the name, International X-ray and Radium Protection Committee (IXRPC) in 1928 when it was formed as a committee of the Second International Congress of Radiology, ICR.[1] This Committee operated with seven members for nine years until 1937 and during this period formulated recommendations on protection from ionizing radiation that were based on earlier recommendations published by the British X-ray and Radium Protection Committee in 1921. During this period a principal concern was protection of the radiologist and his staff. The International X-ray and Radium Protection Committee of the ICR ceased to function during the Second World War years, 1937–1950, and was reorganized with new members and in most respects as a new organization with the name, International Commission on Radiological Protection, ICRP, in 1950.

During the latter part of the doldrum period of IXRPC, 1943 to 1950, there were many publications dealing with protection from ionizing radiation by health physicists and radiobiologists working on the nuclear weapons programmes at Harwell, England, Chalk River, Canada and the US National Laboratories but most of these were in-house classified reports until a few years after the war ended. A large fraction of the members of the ICRP Main Commission and its Committees in the revival period of ICRP (1950–1960) were associated with these laboratories and were joined on the ICRP by medical doctors from these and other countries.

Through the years the ICRP has served as the principal source of information on risks of exposure to ionizing radiation and since 1950 has provided extensive recommendations that have been of assistance to the countries of the world in setting their radiation protection standards, rules and regulations. Some countries have accepted the ICRP recommendations without question as though they were Gospel truth or infallible. Perhaps in most cases they were wise in this reliance, but in some respects I believe ICRP has not met their expectations or justified unqualified acceptance of its recommendations. For this reason I believe it might be helpful to look at what we might consider have been some of the shortcomings of the ICRP, and call attention in a constructive vein to some cases where it was at fault for ignoring radiation exposure problems and to others where it made bad recommendations. Others on this programme are scheduled to provide balance to this topic by enumerating some of the successful accomplishments of ICRP so I leave this discussion to them. Some of the early mistakes of ICRP were reflections of the misconceptions of the science of the times and the fault of ICRP was that perhaps it should have been a bit more ahead of its time. The early publications of the International X-ray and Radium Protection Committee[1] were in various journals, mostly the *British Journal of Radiology*. The ICRP now has in preparation its fiftieth handbook; the first of these, ICRP-1, was published in 1959.

ICRP's greatest mistake in the early period resulted from the false belief of many of its members that low-level exposure is harmless and to many the term tolerance dose connoted a safe dose well below the threshold at which any

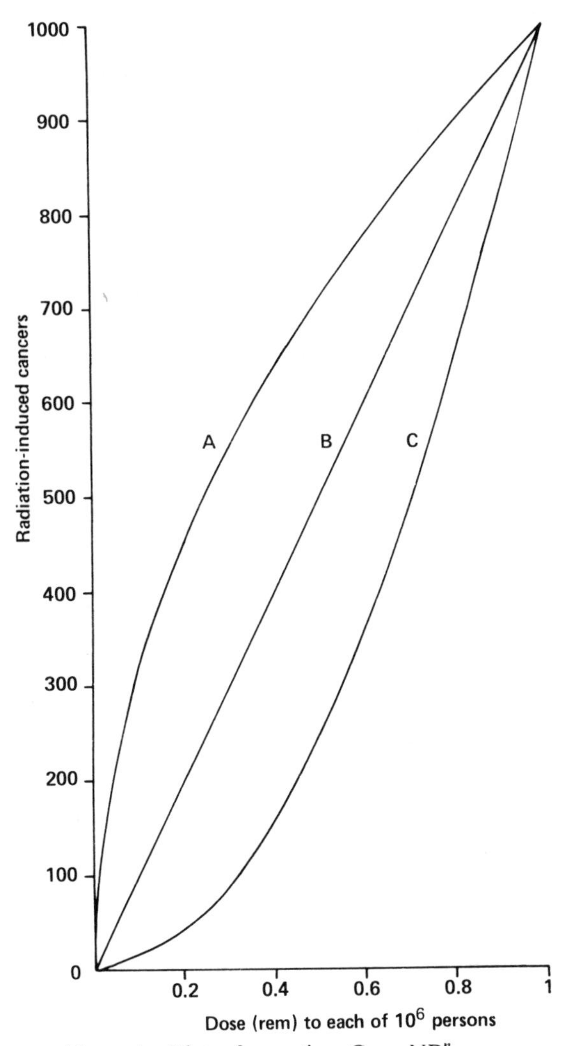

Figure 1 Plot of equation $C = \sigma N D^n$

C = cancer deaths from dose D
D = dose (rems) to each person
N = number of persons = 10^6 in this case
σ = fatal cancers/person rem
$n = 1$ for linear hypothesis, Curve B
$n > 1$ for threshold hypothesis
$= 2$ in Curve C
$n < 1$ for supralinear hypothesis
$= \frac{1}{2}$ in Curve A

harm would ever result. During the years that followed animals and human studies indicated this to be a bad assumption and so the threshold hypothesis was discarded in favour of the linear hypothesis although some of the ICRP publications left the reader with the impression this certainly was a most conservative assumption that without doubt greatly overestimated the risk. Today ICRP is at another crossroad pointing clearly in the direction of the supralinear hypothesis, which not only disclaims a safety factor associated with the application of the linear hypothesis, but asserts that it under-estimates the cancer risk. The three hypotheses can be expressed in a simplified form by the equation, $C = \sigma D^n$, in which C is the radiation-induced cancer incidence or mortality, σ is a constant referred to as the cancer coefficient and D is the dose in rems. If n is greater than 1, we have the threshold hypothesis; if it is equal to 1, we have the linear hypothesis and if it is less than 1, we have the supralinear hypothesis. In many cases of cancer induction, $n = \frac{1}{2}$ or the cancer risk increases with the square root of the dose so that more cancers are produced per rem at low doses than at high doses. Figure 1 illustrates the three hypotheses. Although the ICRP and other agencies frequently are forced to mention results of epidemiological studies where the cancer risk versus the dose of ionizing radiation conforms best with the supralinear theory, they seem to do so reluctantly implying that something most certainly is wrong with such data. For example, the US General Accounting Office[2] formed an expert committee to evaluate the risks of low-level exposure and it concluded, after examining studies of cancer incidence among patients with ankylosing spondylitis who had developed cancer following x-ray medical treatments, that

> Both mixed models tested did much better than the linear model and the unusual square root-cubic model did the best of all. Since the cubic term is negligible at low doses, this last model has a faster than linear growth in leukemia risk for very low doses of X-rays.

The ICRP, UNSCEAR, BEIR-III and other groups are quick to devalue or criticize studies which lend support to the supralinear hypothesis such as the *in utero* studies of Alice Stewart, the tinea capitis studies of B. Modan *et al.* or the Hanford radiation worker study of T. F. Mancuso, A. Stewart and G. Kneale, but they wait for years until they are forced to acknowledge the more obvious and serious flaws in their inspired, irrefutable hallmark, the study of survivors of the atomic bombing of Hiroshima and Nagasaki. They invent all sorts of explanations as to why the former studies are not reliable or admissible in determining σ, the cancer coefficient (σ = cancer deaths per person rem) but fail to recognize the shortcomings of the Japanese study. Some reasons why the study of survivors of bombings of Hiroshima and Nagasaki underestimate cancer risk are:

(1) The total dose estimates were too high, thus σ was underestimated

(2) The neutron dose, especially at Hiroshima, was lower than estimated, thus σ was underestimated
(3) Some evaluations use the low dose group as controls. On the supra-linear hypothesis this could greatly underestimate σ.
(4) It was not a normal population.[3] The bomb survivors had been exposed to fire, blast, deprivation, psychological damage and severe damage to their immune systems so the weaker persons with less resilience died of a variety of common diseases. On the other hand, those of superior stamina or the 'healthy survivors' had a lower than normal death rate five years after the bombing when the epidemiological studies got underway. Thus, as explained by A. M. Stewart,[4] the two effects tended to neutralize each other and when the survivor study began, the death rate appeared normal except for cancer but the population most certainly was far from normal. Thus the cancer incidence was suppressed.
(5) There still is an above normal cancer rate among the survivors and this will continue to increase the value of σ.

In spite of very limited knowledge about the long-term hazards of exposure to low levels of ionizing radiation in the early period (1950–60), I believe in many respects the relative level of excellence as measured by the quality of performance of the Main Commission of ICRP was higher than since then, especially if one takes into account the fact that during this early post-war period ICRP was a pioneer breaking new ground. Certainly much of this credit or blame depends on the stature of the thirteen members comprising the Main Commission of ICRP. I believe it would be difficult to contemplate finding men of less bias and higher qualifications, for example, than Sir Ernest Rock Carling, W. Binks and M. V. Mayneord of the UK, A. J. Cipriani of Canada, R. M. Sievert of Sweden, and G. Failla and H. J. Muller of the US.

Perhaps one of the greatest weaknesses of ICRP stems from its process of nominating and electing members to the Main Commission; however, I must be quick to say it is difficult to think of a perfect solution. The nomination and election processes are flawed because they invite bias and appointment of members who have a conflict of interest and tempt some to make this a lifetime profession assuring them wide political recognition as an authority on radiation protection. In the first place, ICRP functions under the auspices of the International Congress of Radiology, ICR. Possible conflict comes here from the fact that ICRP is set up supposedly to reduce non-beneficial radiation exposure, yet the greater the number of radiation diagnostic procedures and the more routine and assembly-line style in which medical X-rays are administered, the greater the demand for radiology. In many cases this leads to X-rays that are not necessary[5] and to administrative rather than medical requirements for X-rays. On the other hand, when ICRP began in 1928, radiologists comprised the segment of the population with the largest exposure to ionizing radiation and the greatest number with reported radiation injuries. They knew more about its uses, its measurement and its control than any other group. More importantly, the ICR was the first and only international professional organization sufficiently concerned to form such a protection committee and finance its operation. Some of the national society affiliates that comprise the ICR, however, have done more to increase unnecessary patient exposure to X-rays than to minimize it. For example, some of us worked for many years to do away with the mass chest X-ray programme in the schools in the US but we only got negative support of our national radiological societies or the ICR. In this programme buses with photo-fluorometric X-ray equipment would pull up to a school each year and the children were marched through to have a chest X-ray. It would have been better had they instead been branded with a sizzling Texas branding iron because measurements made by my group at Oak Ridge National Laboratory of a number of these devices in use indicated surface doses per X-ray ranging between 2000 and 3000 mR while the average chest X-ray dose at my facility (Oak Ridge National Laboratory) was only 15 mR. Finally, years after the US Surgeon General repeatedly urged a discontinuance of these programmes and after he indicated they had not been finding cases of TB, these programmes were done away with in the US.

Another example of the low priority the American radiological societies and the American College of Radiology, ACR, have given to radiation protection is their reluctance to endorse and failure to make use of the Ten-Day-Rule. This ICRP Rule stated that diagnostic X-rays to the pelvic and abdominal region of women in the child-bearing age should be delayed in most cases and given during the 10-day interval following the beginning of menstruation unless such delay would be harmful to the woman. Dr Muller and I had worked long and hard for ICRP to adopt this Rule and we were delighted when it was adopted by ICRP at the 1962 London meeting. Our delight, however, was short-lived and somewhat impaired when we returned to the US and read in the Bulletin of the ACR that this was a bad Rule and it had been unsuccessfully opposed by two of the members of ICRP, L. S. Taylor and R. S. Stone. It is true this Rule adds to the complexities of operating a radiology department like a factory assembly line and means rescheduling of many X-rays but I believe the unborn child deserves this extra inconvenience and consideration. I have been very disappointed in recent years that ICRP has weakened its stand on the Ten-Day-Rule.

ICRP has not taken full advantage of the findings of Alice Stewart and G. Kneale[6] in their Oxford Studies of *in utero* exposure. It is not to ICRP credit that the permissible likely occupational exposure of pregnant women was decreased from the 1962 value of 1866 mrem[7] only to 1708 mrem[8] in 1977.* I believe this 1708 mrem is far too high. This would correspond to about six

*In 1962 ICRP-6 permitted exposure of 1300 mrem in 13 weeks or an average of 866 mrem in

of the typical pelvimetries delivered during the period of the Oxford Study[9] and ten times the normal risk that the child will die of cancer in early childhood. It is probably true very few mothers would be so calloused as to willingly allow this likely occupational exposure of 1708 mrem to their unborn children but many radiation workers are not aware of the serious warning given us by the Stewart data and certainly many nuclear industries would just as soon this information were not publicized. Even worse, there is nothing in the ICRP recommendations to deter the nuclear industry from allowing the young woman to receive the full 5000 mrem during the two months before pregnancy is recognized (i.e. three times the above risk estimates).

One of the weaknesses of ICRP is in their rules of turnover of membership on the thirteen member ICRP Main Commission. The rules specify not less than two or more than four members shall be changed at each meeting of ICR (every three years) and there is no restriction regarding one's tenure on the Main Commission. Several members have been on the Commission more than twenty years and the average turnover has been 3.7 members every three years. I believe it would be a big improvement to change the rules to require a turnover of not less than four or more than five every three years and have a maximum tenure of nine years. Selection of new members is made every three years by the thirteen member Main Commission from nominations submitted to it by National Delegations to the ICR and by the thirteen member Main Commission members themselves. This has resulted in a self-perpetuated body. I am confident there are several ways in which this election process could be improved. The ICRP has a number of active committees which it appoints from time to time and these usually comprise fifty or more persons in addition to the thirteen members on the Main Commission. Perhaps they too should submit nominations for the Main Commission membership and they, plus the thirteen members of the Main Commission, could vote every three years on the membership. Only Committee members on committees that have been active during the three-year period should have a vote. I am sure such a change would not solve all the membership problems but I believe it would place more qualified persons on the Main Commission to respond to needed or current projects of the Commission. It would more likely result in having certain disciplines properly represented. It would bring in highly qualified scientists from countries seldom, if ever, represented and hopefully it would lessen the chance of special interest groups such as from radiology or the nuclear energy industry having excessive influence. It might lessen the number who have a conflict of interest in reference to current projects of the ICRP or bring to the

---

first two months of pregnancy and 1000 mrem in the last seven months or a total of 1866 mrem. In 1977 ICRP-26 permits exposure of 5000 mrem in a year or 833 mrem in first two months and exposure at the rate of 0.3×5000 mrem/year for the last seven months or a total of $833 + 875 = 1708$ mrem. Both ICRP-6 and ICRP-26 do not actually prohibit the woman receiving exposure of 5000 mrem during the first two months of pregnancy.

top of the agenda new areas where ICRP should operate. I believe there have been two groups excessively represented on the Main Commission of ICRP that have a strong interest in depreciating the harmful effects of low exposure to ionizing radiation. These are persons wishing no restrictions on dose from excessive use of diagnostic X-rays and those with the nuclear establishment (employees of National Laboratories and, with industry and government agencies, responsible for promoting the development of nuclear weapons or supporting nuclear power). These groups need representation but I would like to see them counterbalanced by persons such as, for example, Drs Alice Stewart, J. Rotblat, B. Modan and Frank von Hippel, just to name a few of many who are well qualified.

I believe since ICRP has been considered by many as the most reliable and the ultimate authority on radiation protection for sixty years, its failure to address and try to correct a situation of high, unnecessary radiation exposure must be considered a public disservice. I will mention a few of these faults of omission in the following as typical examples:

(1) In the first period of operation of ICRP, X-ray technicians were instructed in their training classes and in their textbooks[10] to give larger X-ray doses to black people. The General Electric Company's X-ray department recommended in their technique charts for X-ray technologists that they give higher doses to blacks and the textbook, *X-Ray Technology*, by C. A. Jacobs and D. E. Hagen recommended doses to blacks that were higher by 40 to 60 per cent. Why did ICRP remain silent?

(2) The excessive doses delivered in the mass chest X-ray programme to millions of children went on for many years. The dose per X-ray could have been reduced by a factor of 200 by the use of better equipment but dollars were more important than children's lives. Why was ICRP silent on this issue?

(3) During the period of 1960–65 there were many papers published indicating the serious risk of lung cancer among underground uranium miners in the Colorado Plateau region of the US. There were several US Congressional hearings in which many scientists testified—some for reducing the maximum permissible working level in the mines while others urged the level not be reduced. It was no surprise that the US Atomic Energy Commission (USAEC) opposed any reduction in the permissible working level (WL) but I was disappointed that the US Public Health Service and the US Federal Radiation Council (USFRC) joined with the USAEC in opposing any reduction. Table 1[11] indicates the number of men employed in uranium mining in 1954–66 and the high percentage of underground mines operating at very high working levels in 1956–9. Note that for these years only 18 to 28 per cent were operating

Table 1

Estimates of the number of mines producing uranium ore during the calendar year as reported by the industry to the US Bureau of Mines (1954–64) and AEC (1965–66)*[a]

Year	Underground mines	Open pit mines
1954	450	50
1955	600	75
1956	700	100
1957	850	125
1958	850	200
1959	801	165
1960	703	166
1961	497	122
1962	545	139
1963	573	162
1964	471	106
1965	562	74
1966	533	88

Number of men employed in uranium mines[a]

Year	Underground[b] mines	Open pit mines
1954	916	53
1955	1376	293
1956	1770	584
1957	2430	574
1958	2796	1175
1959	3996	1259
1960	4908	1499
1961	4182	1047
1962	4174	1074
1963	3510	886
1964	3249	726
1965	2900	700
1966	2545	359

Estimated distribution of mines by Working Level ranges from 1956 to 1959

Year	Number mines measured	<1.0 WL (%)	1.0–2.9 WL (%)	3.0–10.0 WL (%)	>10.0 WL (%)	Total (%)
1956	108	19	25	33	23	100
1957	158	20	26	28	26	100
1958	53	28	21	36	15	100
1959	237	18	26	28	28	100

[a]Published by the US Federal Radiation Council as report on, *Guidance for Control of Radiation Hazards in Uranium Mining*, Report No. 8 (revised), Sept. 1967.
[b]Excludes above-ground employees who may occasionally go underground.

at a level less than 1 WL ($\sim 10^{-7}$ μCi/cc of RN-222) while ICRP-2 handbook (1959) gave 3×10^{-8} μCi/cc (~ 0.3 WL) as the maximum permissible concentration (MPC) for Rn-222. Figure 2[11] indicates how the cancer risk increased with working level months (WLM). Fortunately after all this discouragement an honest government official turned up in Washington, Secretary of Labor, Mr Wirtz, and he unilaterally set the level at 0.3 WL or $0.3 \times 12 \simeq 4$ WLM (working level months per year). But where was ICRP all this time? It was not until 1977 that ICRP-24,

Radiation Protection in Uranium and Other Mines[12] was published. Surely it should not have taken twenty years for ICRP to decide this was a very serious radiation problem and come to our assistance? One might have expected this to be one of ICRP's first handbooks, warning of the risks of Rn-222 and its daughter products in underground mines. This

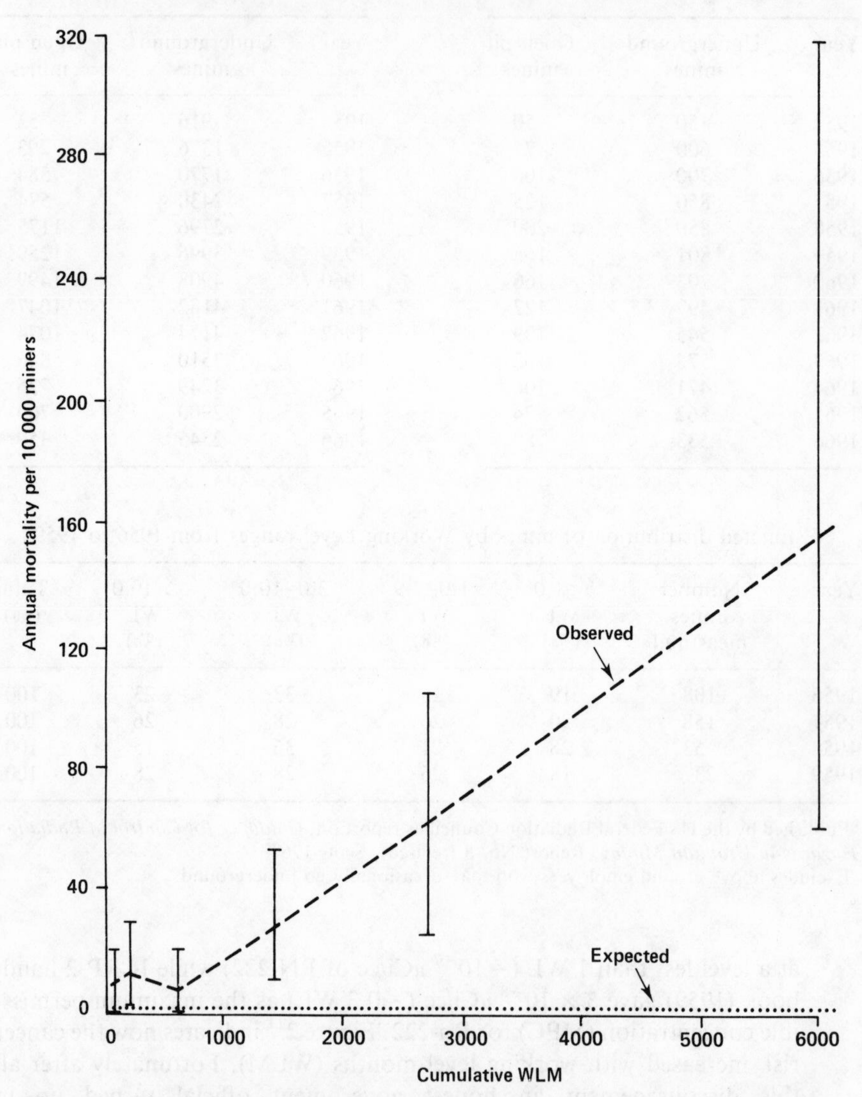

Figure 2 Observed and expected annual lung cancer mortality per 10 000 miners and 95 per cent confidence limits in relation to exposure. (From report of the US Federal Radiation Council titled, '*Guidance for the Control of Radiation Hazards in Uranium Mining*, No. 8 (revised), Sept. 1967.

hazard had first been recognized over 500 years ago when miners in the Schneeberg cobalt mines of Saxony and the Joachimsthal pitchblende miners of Bohemia were dying of a miners' disease, now known as radiation-induced cancer. It is good to have ICRP come in and give support years after a battle is won but it would have been so helpful to have its support to expedite and help the battle earlier.

(4) In the discussion above it was mentioned that ICRP operating under support of the ICR showed great interest in preventing excessive occupational exposure to the radiologists and their staff but dragged its feet in publishing the first comprehensive reports on protection of the patient.

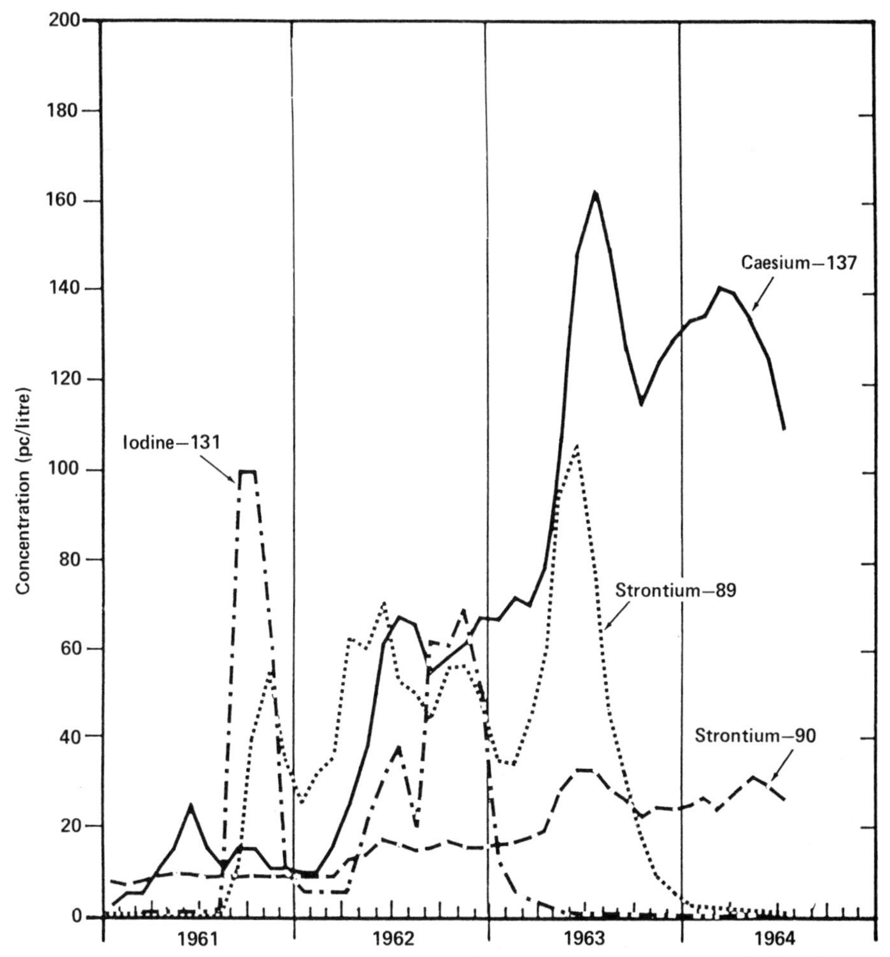

Figure 3 Average concentration of radionuclides in milk samples from Public Health Service pasturized milk network.

These reports finally appeared in 1970[13] and 1982[14] which were 42 and 54 years, respectively, after ICRP began. These are very useful handbooks but far more is needed on this subject.

(5) One of the most frustrating experiences many of us faced over a period of years was the large dose delivered both locally and world wide during atmospheric testing of nuclear weapons. Figure 3 indicates the seriousness of this problem in the US as attested by fallout levels of I-131, Cs-137, Sr-90 and Sr-89 in the pasteurized milk samples collected from 100 sampling stations in the US Public Health Service Network program[15] and Table 2[16] indicates the estimated dose in the 'wet' areas of the US. If one assumes there were 150 million people in the wet areas receiving the whole-body dose of 130 mrem over 70 years and that three quarters of this dose is effective, this corresponds to $\sim 15\,000$ cancer deaths when using a cancer dose coefficient of $\sigma = 10^{-3}$ cancer deaths per rem (i.e. $0.130 \times 150 \times 10^6 \times 3/4 \times 10^{-3} = 14\,625$). Those conducting these tests tried to make the problem seem small, first by using a cancer coefficient σ that was too small by an order of magnitude, i.e. $\sigma = 10^{-4}$, by comparing the dose with that from natural background,

Table 2 Estimated radiation 70-year dose commitment in the wet areas of the United States from nuclear weapon testing in 1962 and from all testing through 1962[a]

Tissue or organ	70-year dose commitment from 1962 testing (mrem)	70-year dose commitment from all testing through 1962 (mrem)
Whole body and reproductive cells		
Caesium-137 external	10	
Caesium-137 internal	10	
Short-lived nuclides	18	
Carbon-14	18	
Total	56	130
Bone		
Strontium-90	180	
Strontium-89	39	
Whole body	56	
Total	275	465
Bone marrow		
Strontium-90	60	
Strontium-89	13	
Whole body	56	
Total	129	215

[a]Values given in report, *Estimate and Evaluation of Fallout in the United States from Nuclear Weapons Testing Conducted through 1962*, Federal Radiation Council, Report No. 4, May 1963.

i.e. ~ 100 mrem per year, and by comparisons with the natural cancer death rate of 20 per cent. In other words, 15 000 cancers on top of 30 million was considered 'negligible'. To me this is absurd. It is like telling a mother whose child is dying of radiation induced cancer not to worry because 30 million other people in the wet area of the US will die naturally of cancer.

When a nation or an industry decides to go ahead with a programme that costs lives, I consider it has reached a conclusion on the value of a human life. Some years ago our Nuclear Regulatory Commission, NRC, made this decision when it set the value of a rem at \$1000, i.e. industry was justified in spending \$1000 to prevent one person-rem. If $\sigma = 10^{-3}$ cancer deaths per rem, this corresponds to $\$1000 \div 10^{-3}$ = a million dollars per life. Others have set the value of a life much lower. For example, two members of the ICRP Main Commission, Mr H. J. Dunster[17] and Dr A. S. McLean (Dr A. S. McLean was a member of the Main Commission of ICRP from 1973 to 1977 and Mr H. J. Dunster has been a member since 1977), published the value of 10 to 25 dollars per man-rem or \$10 000 to \$25 000 per life.

The average world-wide annual whole-body dose commitment from weapons fallout is about 5 mrem per year[18] to the year 2000. This will cause a projected 750 000 cancer deaths (5×10^{-3} rem $\times 10^{-3}$c/person rem $\times 5 \times 10^{9}$ persons $\times 3/4 \times$ effective lifespan of 40 y = 750 000 cancer deaths from 1960 to the year 2000). Again I consider this very significant even though it is only 0.075 per cent of the total cancer deaths in the world population. The fact that other agencies such as the United Nations Scientific Committee on Atomic Radiation (UNSCEAR) addressed the fallout question in no wise obviates the obligation of ICRP, the recognized world authority on effects of radiation on man, from letting its voice be heard by taking up this issue and doing what it can to stop the deaths of 750 000 people. I regret ICRP was silent on this issue.

As I view it, the question of whether an organization like ICRP should set a higher value on a human life or whether it should go against the politically expedient stream, stick its neck out and ask for trouble by trying to put a stop to testing of nuclear weapons in the atmosphere is one of morality. Now an even greater question stands out. Should ICRP, like Physicians for Social Responsibility or the International Physicians for Prevention of Nuclear War, join the battle and help prevent the horrible suffering following World War III in considerable measure due to mass radiation of the world population? Perhaps this question will be answered with the word NO or in the typical way government agencies resolve issues. I quote from a letter uncovered via the US Freedom of Information Act. It was from Dr P. C. Tompkins, then Executive Director of the US Federal Radiation Council to Dr Haworth, Chairman of the US Atomic Energy Commission, dated 25 September 1962. It states in part,

> If any reasonable agreement on this subject can be reached among Agencies, the basic approach to the report would be to start with a simple, straightforward statement of conclusions. It would then be a straightforward matter to select the key scientific consultants whose opinions should be sought in order to substantiate the validity of the conclusions or recommended appropriate modifications.

I hope ICRP will not operate like Government Agencies! I do believe, however, this was the basis on which the US Atomic Energy Commission decided the so-called particle problem (high dose near a small radioactive particle) was not a problem. I was never satisfied that the decision makers took proper account of studies such as those of H. Lisco *et al.*[19] where they observed a high incidence of cancer at the sites of injection of Pu-239 and other radionuclides under the skin of animals.

(6) An ICRP report on major radiation accidents, problems encountered and in what manner they were handled is long overdue. In 1984 ICRP published[20] a report on major radiation accidents but it is very brief, superficial and fails to address the emergency situations that have been experienced in major radiation accidents in many places. There have been four fatal criticality accidents in the US (Los Alamos, NM on 21 August 1945, 21 May 1946 and 30 Dec, 1958 and Wood River Junction, RI on 24 July 1964) and the SL-1 Reactor explosion at Idaho Falls, ID, 3 January 1961. One person was killed in each of the four criticality accidents and three were killed in the SL-1 accident. Only this SL-1 accident resulted in environmental contamination but because of its isolation in a desert environment, contamination beyond the plant-site and into the public domain was minimal. There were several bad accidents at the Rocky Flats Plant[22] not far from Denver, CO and these resulted in widespread environmental contamination from Pu-239. Recently information was uncovered indicating that on 2 December 1949, 5500 Ci of I-131 were released at the Hanford, WA plant to test meteorological patterns and adequacy of instruments in case fission products were used as an adjunct to chemical warfare. Also there had been hundreds of thousands of Curies released into the air and into the Columbia River during operations of the plant in weapons production during the war. The Savannah River, SC plant[23] had a number of releases of radioactive material into the environment. All of these accidents could provide a wealth of useful information on what to do and not to do and this information is just waiting for an organization like ICRP to bring it together. There have been also many non-fatal criticality accidents like the one at Y-12 in Oak Ridge, TN on 16 June 1958 in which five workers

got over 200 rem doses and there was the mild explosion at Mol, Belgium on 30 December 1965 where a worker received 550 rem. The Vinca, Yugoslavia accident on 15 October 1958 resulted in four persons receiving over 400 rem and one of these died after 32 days. I consider it a shame that much valuable information about these accidents has not been brought together and put into print. At present it only lingers in the minds of a few persons still living.

So far as I know until Chernobyl there had been only three major reactor accidents (SL-1 reactor, 3 January 1961, Windscale reactor No. 1 on 8 October 1957 and Three Mile Island Reactor no. 2 on 28 March 1979). There was a massive explosion[24,25] apparently in improperly buried nuclear waste in the Ural Mountains in 1957, but I have only fragmentary information on this. Possibly the Russian member on the ICRP Commission could provide details of this accident for this long awaited ICRP handbook. Some of the information I have in mind could have been of assistance at the time of the Chernobyl accident. In the following I list a few personal experiences that suggest the kind of information that should be given in this ICRP handbook that ought to have been written many years ago. They are as follows:

(a) At the SL-1 accident the men who ran up the stairs of the reactor building on a rescue attempt were in a thick cloud of dust and in radiation field of hundreds of rem/h. They certainly received very large doses to the nasopharyngeal and tracheobronchial regions of the lungs and to the gastrointestinal tract from the inhalation of large dust particles. It is unfortunate that faecal as well as urine blood and sputum samples were not collected and analysed after this exposure. The autopsy data on the three bodies (of those caught in the blast and eventually recovered from the debris in the reactor building) provided extremely useful information on the nature and cause of the accident.

(b) Following the explosion of the chemical process tank at Oak Ridge National Laboratory in 1965, it was very important that we approach the scene of the accident with operating neutron dose meters in hand. Although not known at the time, it turned out later there was enough plutonium in this tank for many critical assemblies. At any moment the liquid could have settled into a critical configuration.

In this same explosion a large amount of plutonium was blown out over an adjacent building and onto a road. Within four hours after the accident we had tarred the road and sprayed the adjacent building with paint. Later the road was taken up piece-by-piece, placed in plastic containers and sent to the official burial ground. The building later was disassembled piece-by-piece, placed in burial containers and properly buried. I have seen pictures from Chernobyl where they are washing down the buildings and roads with water. This is opposite of what I would recommend except for contamination by short-lived radionuclides.

(c) In the Y-12 accident mentioned above I had all sorts of meters as I entered this building and homed-in on the criticality assembly with my operating instruments in hand but I failed to have with me a much needed instrument—a simple flashlight to see and to read the meters—for the electricity was now cut off in this labyrinth of a windowless building. In major accidents, important but simple things often are lacking. For example, gasoline pumps may not operate because the electrical power is knocked out. After the Y-12 accident I had the plant doctor collect 5 cc of blood from each of the highly exposed persons, mix it with heparin to prevent coagulation, and we measured the P-32 and Na-24 in the blood to determine the fast and thermal neutron dose (i.e. $^{32}S(n,p)^{32}P$ and $^{23}Na(n,\gamma)^{24}Na$). Differential blood counts were made from time to time and studies of chromosal aberrations were carried out.

(d) When I visited Windscale a couple of days after the accident, I was told of two major problems: (1) they did not get their light aircraft airborne for aerial surveys soon enough and (2) utter confusion at times could have been avoided if they had had a well equipped communication centre ready and waiting for them at the time of the accident. Neither of these two things were available at the Three Mile Accident or the Chernobyl accident. Maybe if ICRP will prepare this handbook on accidents, emergency personnel will be better prepared for the next major reactor accident?

(e) Perhaps ICRP in this proposed handbook could give guidance on how to put out a fire in graphite, uranium or zirconium? We had a fire at the back of our graphite reactor at Oak Ridge National Laboratory about the time we had a visit from Sir John Cockcroft in the late 1940s and he was impressed with the necessity of filters in the cooling air from a reactor before the cloud of dust and smoke went up the stack. But alas when he returned to the UK, the Windscale stack was already half built. But nothing could stop a great scientist. An immense filter house was built halfway up the stack. This became known as 'Cockcroft's folly' but it partly saved the day during the graphite fire at Windscale on 8 October 1957.

Early at Oak Ridge and at Windscale and recently at Chernobyl water was used with much trepidation to extinguish the fires but it put them out. However, at Chernobyl the water probably reacted with the hot metals and graphite to produce large amounts of hydrogen. While I was at Windscale during the time of the accident early in October 1957, I was puzzled that even though the filters near the top of the stack were saturated with water, they had held up most of the Sr-89 and Sr-90, Cs-134 and Cs-137 and the I-131, I-132, I-133, I-134 and I-135. How could this be? Then I was told that dense fumes from Bi-209 which was stored in the burning part of the reactor had acted as condensation nuclei and made even water-saturated filters relatively efficient. I knew of course that the Bi-209 was in the reactor to produce Po-210 for the neutron trigger then used in our atomic bombs,* so I kept this a dark secret in my mind until a few years ago when information was declassified and released that Po-210 was one of the Windscale fallout products in 1957. Incidentally, about ten years ago Dr A. Stewart and I had just given lectures at a meeting in London and in the question period H. J. Dunster (now a member of ICRP) criticized Dr Stewart for having said an alpha-emitter was discharged with the fission products. In the discussion I tried in a weak way to come to Dr Stewart's rescue, but my lips were sealed because of security. Po-210, an alpha-emitter, was discharged along with the beta- and gamma-emitting fission products during the Windscale accident.

(7) One subject which I believe should be carefully followed by ICRP and on which it could make very useful recommendations, for example, is that of the person-rem per year at the various nuclear plants. This should be addressed both in terms of person-rem per year per plant and person-rem per year per 1000 MWe. It is astonishing to note that some nuclear power facilities consistently have a better record than others in this regard by more than an order of magnitude.[26,27]

(8) A final example of where ICRP, in my estimation, has been somewhat negligent is in meeting the need of an in-depth treatment of the environmental releases of radionuclides of greatest concern in the nuclear industry. Here we think of H-3, C-14, Sr-89 and -90, I-131, Cs-134 and -137, noble gas etc. Such a publication might help to answer many recurrent questions such as: What are the genetic risks of these radionuclides? Was ICRP justified in reducing the quality factor of the low energy beta radiation of H-3 from 1.8 to 1.0 when theory suggests the value of 2 is more appropriate? Why do some nuclear power plants discharge routinely into the environment a hundred times the curies of fission products released by the average power plant? Should additional efforts be made to reduce the large routine release of noble gases and H-3 by a nuclear power plant? Should power plants monitor the release of C-14?

In the foregoing I have discussed what I consider are some of the weaknesses of omission of ICRP and given a few typical examples. Perhaps there are as many weaknesses of commission by ICRP but in the following I will discuss only one, namely the ICRP-26 handbook and how it has resulted in an increase

*$^{209}Bi(n + \gamma)^{210}Bi \xrightarrow[5d]{\beta} {}^{210}Po$ (138d) and $^{210}Po(\alpha, Be)$ neutrons

in values of maximum permissible concentration (MPC) for many of the more common and more dangerous radionuclides such as Sr-89 and Sr-90, I-131 and Pu-238, Pu-239 and Pu-240. The values now recommended by ICRP are higher than we developed in 1959 for ICRP-2 when I was chairman of the Internal Dose Committee of ICRP. This increase might be justified were the risk of radiation-induced cancer much less than we perceived it to be almost thirty years ago but just the contrary is the case; today the cancer coefficient is known to be at least an order of magnitude greater than it was perceived to be in 1959.

During the last few years that I was an active member of the Main Commission of ICRP, we discussed an inconsistency in our basic internal dose standard, namely, the values of MPC were based on concentrations in air, water and food that at the end of an occupational exposure period of fifty years would result in dose rates of 5 rem/y to total body, gonads and active (red) bone marrow, 30 rem/y to bone, thyroid and skin and 15 rem/y to any other body organs that were the critical body organs (usually the organ with the greatest concentration of the radionuclide). In short, our Internal Dose Committee was criticized for using the same dose rate limit for gonads and active marrow as for whole body because, were the whole body exposed to 5 rem/y, the gonads and active bone marrow also would be exposed to 5 rem/y. Partial-body exposure was known to be less harmful than whole-body exposure so the permissible dose rate of the whole body should not be the same as that to the gonads and active bone marrow. It seemed to me the solution was very simple, namely reduce the whole-body dose rate to 2.5 rem/year. However, some members felt this would be a hardship to the nuclear industry and we should keep the limiting whole body dose at 5 rem/y for the nuclear worker exposure both to internal and external sources of radiation. I took the view that the external dose limit of 5 rem/y as well as the internal dose limit was too high and both should be reduced to be more in conformance with our realization that the cancer risk from radiation was greater than we thought it to be when these limits were first set. Unfortunately, in 1977, some years after I had been moved to the status of an emeritus member, ICRP-26 was adopted in which the limiting dose rates after fifty years of occupational exposure were set at 5 rem/y to whole body, 20 rem/y to gonads (an increase by a factor of 4), 42 rem/y to active bone marrow (an increase by a factor of 8.3), 42 rem/y to lungs (an increase by a factor of 2.8), 50 rem/y to thyroid and bone surfaces (an increase by a factor of 1.7) and 33 rem/y to breasts (an increase by a factor of 2.2).

While I was still an active member of ICRP (not yet moved to the status of an emeritus member) we also discussed the possibility that with the coming of the computer age we should be more sophisticated and calculate the MPC not just on the basis of the dose to a critical body organ from the radionuclide that was in this organ but also from what was in all the body organs as they irradiated the critical body organ (now called the target organ). I am pleased to say that ICRP-26 and ICRP-30 followed this suggestion and this has resulted in an improvement over ICRP-2. Had it not been for this latter change, all MPC values now provided by ICRP-26 and ICRP-30 would be increased. Table 3 indicates some of these changes for a few of the important radionuclides. I do not see justifications for any of these values being greater than they were in 1959 when ICRP-2 was published. It should be mentioned also that in ICRP-30[28] the term (MPC) in air has been changed to (DAC) or Derived Air Concentration and values of (MPC) for water for some unknown reason are no longer given. Instead, values of (ALI) or Annual Limit on Intake are given. Changing from the curie to the becquerel (= 1 disintegration per second) was bad enough (because we already have the unit hertz with the same dimensions) but now the ALI makes it difficult for us in the US to make direct use of ICRP-30 since we must keep in mind that

$$(\mathrm{MPC})_w = 10^{-10}\mathrm{ALI}\ (\mathrm{Bq})\ \mu\mathrm{Ci/cc\ of\ water}$$

$$(\mathrm{MPC})_a = \frac{\mathrm{DAC}(\mathrm{Bq/m^3})}{3.7 \times 10^{10}}\ \mu\mathrm{Ci/cc\ of\ air}$$

In this list of 46 radionuclides in Table 3 there are 36 cases for radionuclides in air where the $(\mathrm{MPC})_a$ has been increased and ten cases where these is a decrease. For radionuclides in water (or most foods) there are 35 cases where the $(\mathrm{MPC})_w$ is increased and 19 cases where they are lower. I believe all values should be lower. This same ratio of increased to decreased MPC values is maintained approximately in the other 200 radionuclides listed in ICRP-2 but not shown in Table 3.

It was of interest to me to note that when ICRP set up a table of weighting factors in ICRP-26[8] to obtain the new limiting dose rates (as given above) following fifty years of occupational exposure (or the limiting committed dose from a year's exposure) to eliminate the above-mentioned long recognized inconsistency, it did the equivalent of jumping from the frying-pan into the fire, i.e. it made an even more inconsistent move. Using their chosen weighting factors the limiting dose rate for thyroid and bone surfaces turned out to be 167 rem/y. Although all limiting occupational exposure levels are set to limit stochastic* damage and in particular radiation-induced cancer, ICRP was now faced with the fact that 167 rem/y could be expected to result in non-stochastic forms of damage among radiation workers. This of course could not be tolerated so ICRP reached up and adopted the figure of 50 rem/y out of thin air with no justification in terms of cancer induction.

*Stochastic forms of damage are those like cancer that have no threshold and the damage once it develops is not a function of the magnitude of the dose that caused it. Examples are cancer and genetic mutations. Non-stochastic forms of damage do not show up unless a threshold dose is exceeded and the larger the dose that caused them the more severe the effect. Examples are radiation-induced erythema, epilation, cataracts and radiation sickness.

Table 3 Comparison of ICRP-30 with ICRP-2 values

Radio-nuclide	Half-life	ICRP-30 DAC-(Bq/m^3) MPC_a ($\mu Ci/cc)_a$	ICRP-2 MPC_a ($\mu Ci/cc)_a$	ICRP-30[a] ALI (Bq) MPC_w ($\mu Ci/cc)_w$	ICRP-2 MPC_w ($\mu Ci/cc)_w$
H-3(H_2O)	12.26y	8×10^5		3×10^9	
		(2.2×10^{-5})	(5×10^{-6})	(0.3)	(0.1)
C-14(CO_2)	5730y	3×10^6		9×10^7	
		(8.1×10^{-5})	(4×10^{-6})	(9×10^{-3})	(0.02)
Na-24	14.96h	8×10^4		10^8	
		(2.2×10^{-6})	(10^{-7})	(0.01)	(8×10^{-4})
P-32	14.28d	6×10^3		2×10^7	
		(1.6×10^{-7})	(7×10^{-8})	(2×10^{-3})	(5×10^{-4})
S-35	87.9h	3×10^4		4×10^8, 2×10^8	
		(8.1×10^{-7})	(3×10^{-7})	(0.04, 0.02)	(2×10^{-3})
Cl-36	3.1×10^5y	4×10^3		6×10^7	
		(1.1×10^{-7})	$(2 \times 10^{-8}$	(6×10^{-3})	(2×10^{-3})
Ca-45	165d	1×10^4		6×10^7	
		(2.7×10^{-7})	(3×10^{-8})	(6×10^{-3})	(3×10^{-4})
Cr-51	27.8d	3×10^5		10^9	
		(8.1×10^{-6})	(2×10^{-6})	0.1	(0.05)
Mn-54	303d	1×10^4		7×10^7	
		(2.7×10^{-7})	(4×10^{-8})	(7×10^{-3})	(3×10^{-3})
Fe-55	2.6y	6×10^4		3×10^8	
		(1.6×10^{-1})	(9×10^{-7})	(0.03)	(0.02)
Fe-59	45.6d	8×10^3		3×10^7	
		(2.2×10^{-7})	(5×10^{-8})	(3×10^{-3})	(2×10^{-3})
Co-58	71.3d	1×10^4		6×10^7, 5×10^7	
		(2.7×10^{-7})	(5×10^{-8})	$(6 \times 10^{-3}$, $5 \times 10^{-3})$	(3×10^{-3})
Co-60	5.26y	500		2×10^7, 7×10^6	
		(1.4×10^{-8})	(9×10^{-9})	$(2 \times 10^{-3}$, $7 \times 10^{-4})$	(10^{-3})
Ni-59	8×10^4y	3×10^4		9×10^8	
		(8.1×10^{-7})	(5×10^{-7})	(0.09)	(6×10^{-3})
Cu-64	12.8h	3×10^5		4×10^8	
		(8.1×10^{-6})	(10^{-6})	(0.04)	(6×10^{-3})
Zn-65	245d	4×10^3		10^7	
		(1.1×10^{-7})	(6×10^{-8})	(10^{-3})	(3×10^{-3})
Sr-89	52.7d	2×10^3		2×10^7	
		(5.4×10^{-8})	(3×10^{-8})	(2×10^{-3})	(3×10^{-4})
Sr-90	27.7y	60	$[10^{-9}]$[b]	10^6, 2×10^7	$[10^{-5}]$[b]
		(1.6×10^{-9})	(3×10^{-10})	$(10^{-4}$, $2 \times 10^{-3})$	(4×10^{-6})
Mo-99	66.7h	2×10^4		6×10^7, 4×10^7	
		(5.4×10^{-7})	(2×10^{-7})	$(6 \times 10^{-3}$, $4 \times 10^{-3})$	(10^{-3})
Ru-106	368d	200		7×10^6	
		(5.4×10^{-9})	(6×10^{-9})	(7×10^{-4})	(3×10^{-4})
Te-127m	109d	4×10^3		2×10^7, 10^7	
		(1.1×10^{-7})	(4×10^{-8})	$(2 \times 10^{-3}$, $10^{-3})$	(2×10^{-3})
I-126	2.6h	500		8×10^5	
		(1.4×10^{-8})	(8×10^{-9})	(8×10^{-5})	(5×10^{-5})
I-129	1.7×10^7y	100		2×10^5	
		(2.7×10^{-9})	(2×10^{-9})	(2×10^{-5})	(10^{-5})
I-131	8.05d	700		10^6	
		$(1.9 \times 10^{-8}$	(9×10^{-9})	(10^{-4})	(6×10^{-5})

Table 3 *Continued*

Radio-nuclide	Half-life	ICRP-30 DAC-(Bq/m^3) MPC_a $(\mu Ci/cc)_a$	ICRP-2 MPC_a $(\mu Ci/cc)_a$	ICRP-30[a] ALI (Bq) MPC_w $(\mu Ci/cc)_w$	ICRP-2 MPC_w $(\mu Ci/cc)_w$
I-132	2.26h	1×10^5		10^8	
		(2.7×10^{-6})	(2×10^{-7})	(0.01)	(2×10^{-3})
I-133	20.3h	4×10^3		5×10^6	
		(1.1×10^{-7})	(3×10^{-8})	(5×10^{-4})	(2×10^{-4})
I-135	6.68h	2×10^4		3×10^7	
		(5.4×10^{-7})	(10^{-7})	(3×10^{-3})	(7×10^{-4})
Xe-133	5.27d	4×10^6		—	
		(1.1×10^{-4})	(10^{-5})	—	—
Cs-134	2.046y	2×10^3		3×10^6	
		(5.4×10^{-8})	(10^{-8})	(3×10^{-4})	(3×10^{-4})
Cs-137	30.0y	2×10^3		4×10^6	
		(5.4×10^{-8})	(10^{-8})	(4×10^{-4})	(4×10^{-4})
Ba-140	12.8d	2×10^4		2×10^7	
		(5.4×10^{-7})	(4×10^{-8})	(2×10^{-3})	(7×10^{-4})
Ce-144	284d	200		8×10^6	
		(5.4×10^{-9})	(6×10^{-9})	(8×10^{-4})	(3×10^{-4})
Ir-192	74.2d	3×10^3		4×10^7	
		(8.1×10^{-8})	(3×10^{-8})	(4×10^{-3})	(10^{-3})
Po-210	138.4d	10		10^5	
		(2.7×10^{-10})	(2×10^{-10})	10^{-5}	(2×10^{-5})
Ra-226	1602y	10		7×10^4	
		(2.7×10^{-10})	(3×10^{-11})	(7×10^{-6})	(4×10^{-7})
Th-232	1.41×10^{10}y	4×10^{-2}		3×10^4	
		(1.1×10^{-12})	(2×10^{-12})	(3×10^{-6})	(5×10^{-5})
U-234	2.47×10^5y	6×10^{-1}		4×10^5, 7×10^6	
		(1.6×10^{-11})	(10^{-10})	$(4 \times 10^{-5}$, $7 \times 10^{-4})$	(9×10^{-4})
U-235	7.1×10^8y	6×10^{-1}		5×10^5, 7×10^6	
		(1.6×10^{-11})	(10^{-10})	$(5 \times 10^{-5}$, $7 \times 10^{-4})$	(8×10^{-4})
U-238	4.51×10^9y	7×10^{-1}		5×10^5, 8×10^6	
		(1.9×10^{-11})	(7×10^{-11})	$(5 \times 10^{-5}$, $8 \times 10^{-4})$	(10^{-3})
Np-237	2.14×10^6y	9×10^{-2}		3×10^3	
		(2.4×10^{-12})	(4×10^{-12})	(3×10^{-7})	(9×10^{-5})
Pu-238	86.4y	3×10^{-1}		3×10^5, 3×10^6	
		(8.1×10^{-12})	(2×10^{-12})	$(3 \times 10^{-5}$, $3 \times 10^{-4})$	(10^{-4})
Pu-239	24390y	2×10^{-1}		2×10^5, 2×10^6	
		(5.4×10^{-12})	(2×10^{-12})	$(2 \times 10^{-5}$, $2 \times 10^{-4})$	(10^{-4})
Pu-240	6580y	2×10^{-1}		2×10^5, 2×10^6	
		(5.4×10^{-12})	(2×10^{-12})	$(2 \times 10^{-5}$, $2 \times 10^{-4})$	(10^{-4})
Am-241	458y	8×10^{-2}		5×10^4	
		(2.2×10^{-12})	(6×10^{-12})	$(5 \times 10^{-6}$	(10^{-4})
Am-243	7.95×10^3y	8×10^{-2}		5×10^4	
		(2.2×10^{-12})	(6×10^{-12})	(5×10^{-6})	(10^{-4})
Cm-244	17.6y	2×10^{-1}		9×10^4	
		(5.4×10^{-12})	(9×10^{-12})	(9×10^{-6})	(2×10^{-4})

[a] Class W only

[b] Value given in ICRP-6 (1962).

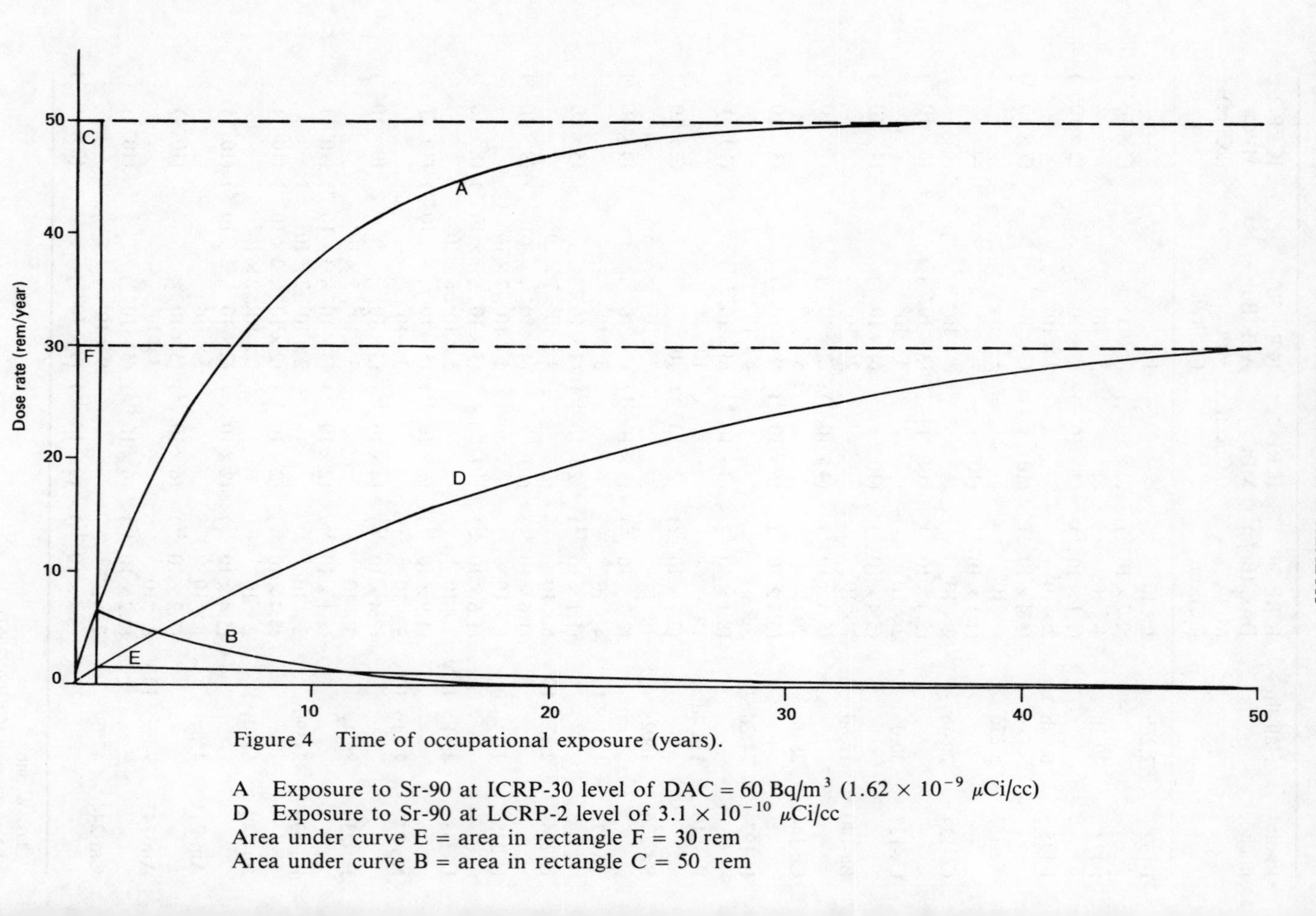

Figure 4 Time of occupational exposure (years).

A Exposure to Sr-90 at ICRP-30 level of DAC = 60 Bq/m^3 (1.62×10^{-9} μCi/cc)
D Exposure to Sr-90 at LCRP-2 level of 3.1×10^{-10} μCi/cc
Area under curve E = area in rectangle F = 30 rem
Area under curve B = area in rectangle C = 50 rem

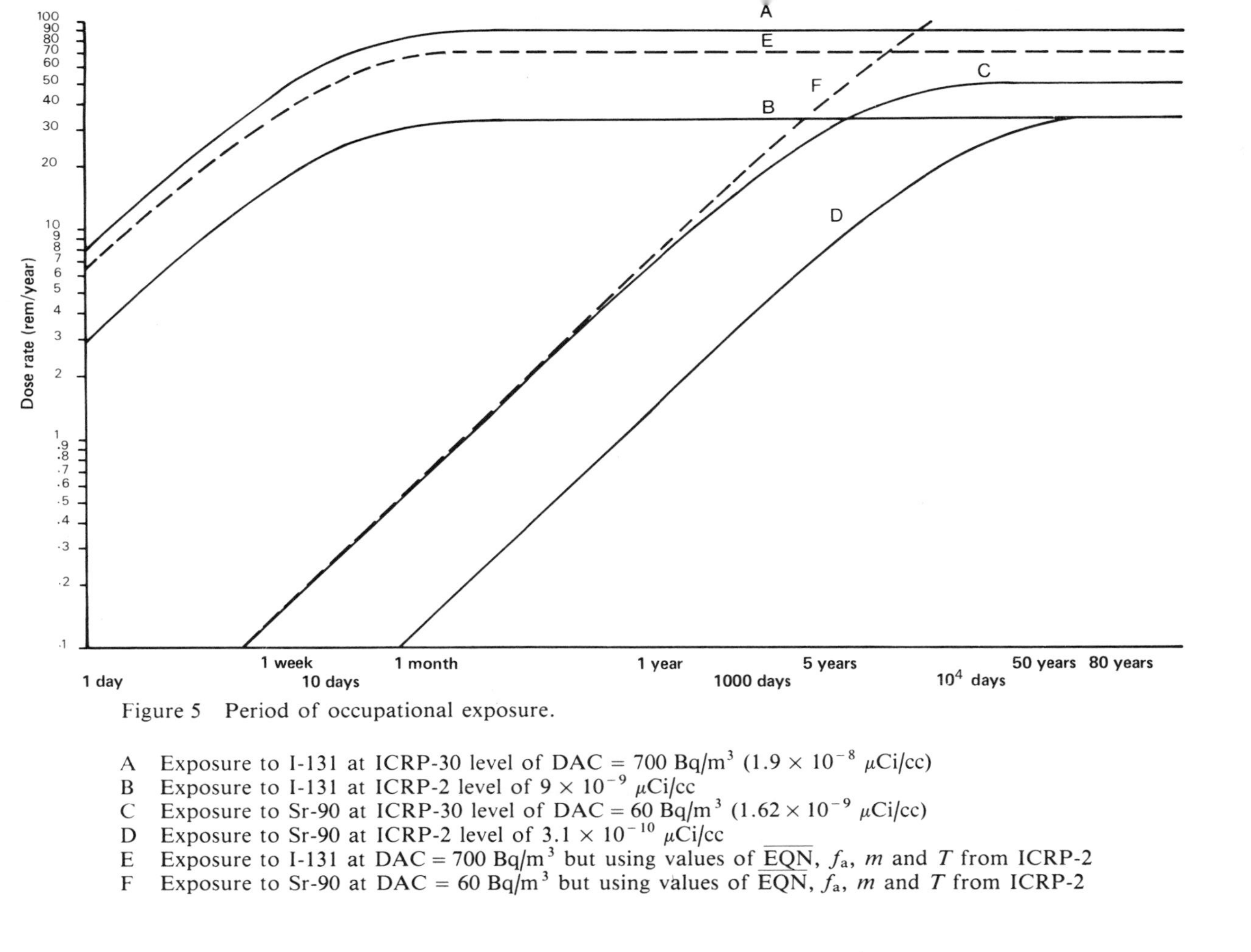

Figure 5 Period of occupational exposure.

A Exposure to I-131 at ICRP-30 level of DAC = 700 Bq/m^3 (1.9×10^{-8} $\mu Ci/cc$)
B Exposure to I-131 at ICRP-2 level of 9×10^{-9} $\mu Ci/cc$
C Exposure to Sr-90 at ICRP-30 level of DAC = 60 Bq/m^3 (1.62×10^{-9} $\mu Ci/cc$)
D Exposure to Sr-90 at ICRP-2 level of 3.1×10^{-10} $\mu Ci/cc$
E Exposure to I-131 at DAC = 700 Bq/m^3 but using values of $\overline{EQN}$, f_a, m and T from ICRP-2
F Exposure to Sr-90 at DAC = 60 Bq/m^3 but using values of $\overline{EQN}$, f_a, m and T from ICRP-2

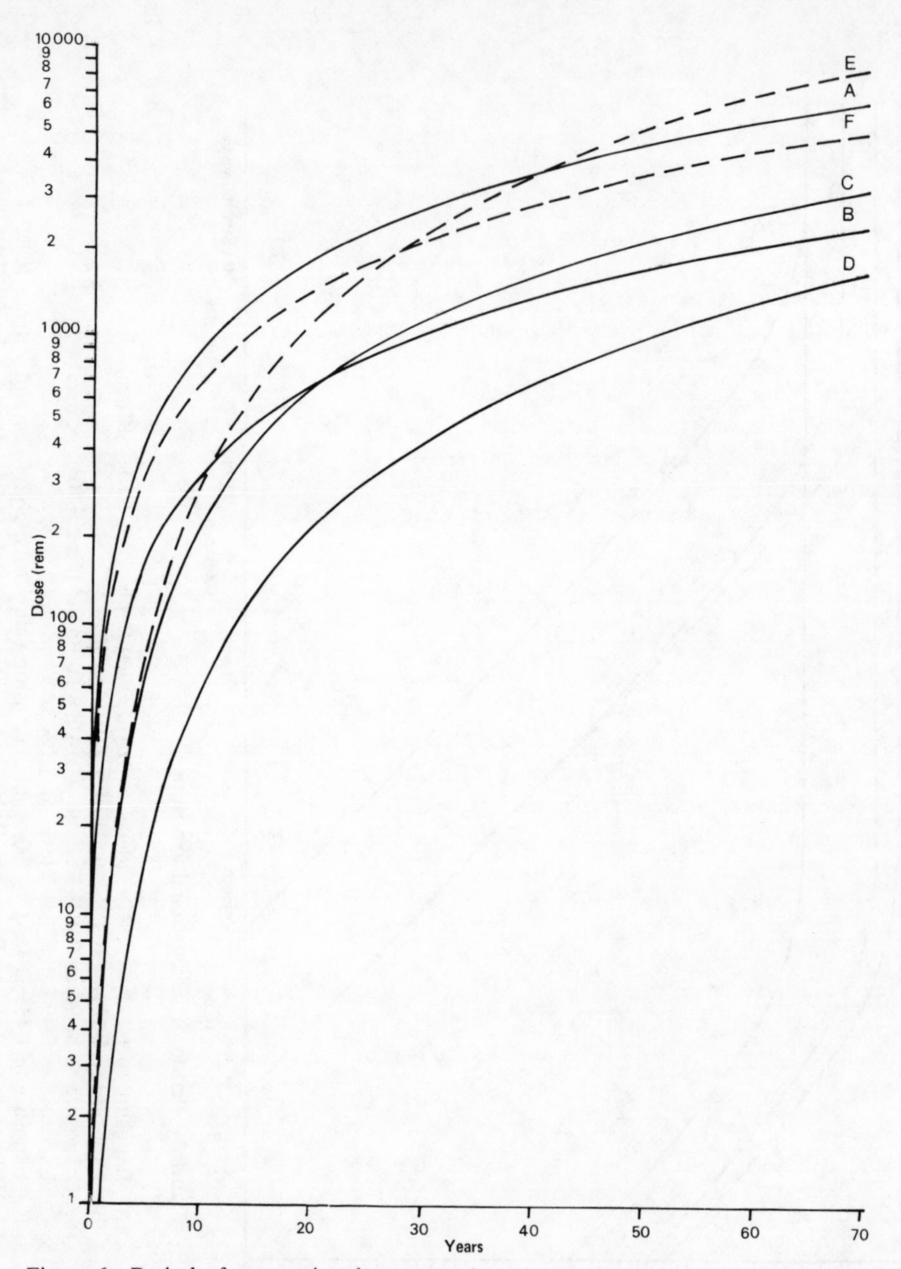

Figure 6 Period of occupational exposure (years).

A Accumulated dose from exposure to I-131 at ICRP-30 level of DAC = 700 Bq/m^3 (1.9×10^{-8} μCi/cc)

In the above it is pointed out that ICRP has now increased the limiting dose rate to all individual body organs except when the whole body itself is the critical body organ, i.e. the radionuclide is distributed rather uniformly in the body such as would be the case for H-3 taken in as HTO. The dose rates listed above are those to which the body organ of the radiation worker would be subjected were he to be exposed at the MPC of one of these radionuclides in air, water or food for fifty years at 40 hours per week and 50 weeks per year. This limiting dose rate, after fifty years as given by ICRP-30 for example, is 50 rem/y to bone surfaces and to the thyroid. Actually in these cases ICRP sets what it calls the limiting committed effective dose equivalent, CEDE, [8,29] during a year at 50 rem. That is, a worker, for example, can take in Sr-90 during a year in any manner provided the dose from this one year intake is no more than 50 rem when the bone surface dose is integrated thereafter over fifty years. In Appendix I it is shown that intake for a work year at the MPC will deliver a critical body organ dose integrated over fifty years that is equal to this CEDE and this CEDE is equal to the dose rate that would be reached in this organ after fifty years of occupational exposure multiplied by one year, i.e. it is numerically equal to the dose rate reached in this organ following fifty years' exposure at the MPC. This is shown in Figure 4 for the case of occupational exposure to Sr-90. Curve A shows the increasing dose rate to bone surfaces of the radiation worker working in a constant work environment 40 hours per week, 50 weeks per year for fifty years when the air concentration is maintained at the present ICRP-30 DAC of 60 Bq/m^3 (1.62×10^{-9} μCi/cc) of Sr-90. In this case I took the ICRP value of $f_a = 0.01$ as the fraction of Sr-90 going to bone surfaces and $m = 120$ g as the mass of these surface tissues. ICRP-30 did not give a separate value for the fraction of Sr-90 going from blood to bone surfaces so I interpreted the 0.01 value to be the product of the fraction to blood by the fraction from blood to bone surfaces. A value was not given in ICRP-30 for the biological half-life so I back calculated to get $T = T_b T_r/(T_b + T_r) = 4.961$ years in order that the fifty-year integration of dose rate from a year's intake would be 50 rem as required by ICRP-30. I used the $\overline{EQN}$ 1.1 MeV per disintegration of Sr-90 plus its daughter Y-90. It should be emphasized that ICRP-2 set $N = 5$ or $\overline{EQN} = 5.5$ MeV per disintegration and since I see no justification for ICRP having set $N = 1$ in ICRP-30, I believe

---

B Accumulated dose from exposure to I-131 at ICRP-2 level of 9×10^{-9} μCi/cc
C Accumulated dose from exposure to Sr-90 at ICRP-30 level of DAC = 60 Bq/m^3 (1.62×10^{-9} μCi/cc)
D Accumulated dose from exposure to Sr-90 at ICRP-2 level of 3.1×10^{-10} μCi/cc
E Accumulated dose from exposure to Sr-90 at DAC = 60 Bq/m^3 but using values of $\overline{EQN}$, f_a, m and T as given in ICRP-2
F Accumulated dose from exposure to I-131 at DAC = 700 Bq/m^3 but using values of $\overline{EQN}$, f_a, m and T as given in ICRP-2

the dose rate values of curve A in Figure 4 are under-estimated at least by a factor of 5. The equations for curves A and D are derived in Appendix II. It is noted that both curves pass through the dose rate limit at fifty years—Curve A at 50 rem/year and Curve D at 30 rem/year. Curve A with a shorter half-life ($T = 4.96$y) reaches its equilibrium at 50 rem/y in about forty years whereas Curve D with a longer half-life ($T = 17.53$y) passes through 30 rem/y (at 86 per cent of equilibrium) at fifty years but as shown by Curve D in Figure 5 it would not reach its equilibrium level of 34.8 rem/y until about 150 years. As indicated by Curve C in Figure 6, exposure at the new DAC of 60 Bq/m^3 (1.62×10^{-9} μCi/cc) for Sr-90 for fifty years would result in an average bone surface dose of 2140 rem but, as indicated by Curve E, when applying the values of $\overline{EQN}$ m and T as given in ICRP-2 this would result in a total bone dose of 5130 rem. Figures 5 and 6 indicate similar increases in thyroid dose rate and dose for I-131 except that the equilibrium dose rate in Curve A of Figure 5 is 91 rem/y instead of the ICRP-30 limit of 50 rem/y. No reason is given in ICRP-30 for this apparent discrepancy. Figure 6 indicates the 50-year accumulated dose by ICRP-30 specifications is 4570 rem (Curve A) and by ICRP-2 specifications is 1680 rem (Curve B). I see no justification for any of these increases.

CONCLUSION

In conclusion I wish to re-emphasize that I have limited this discussion to criticisms of the work of the Main Commission of ICRP in its accomplishments and lack of them over the past sixty years. In this I am pointing the finger at myself as well as to others because I was one of this thirteen member body for about twenty years. Some of the committees of ICRP have done an excellent job. My principal criticisms of the Main Commission are that in many cases it has not responded to important situations of high exposure to ionizing radiation or has been unnecessarily slow in response and it has increased permissible exposure levels at a time when they should have been reduced.

REFERENCES

1. L. S. Taylor, *Radiation Protection Standards*, CRC Press, Cleveland, Ohio, US.
2. Report to the Congress of the United States, *Problems in Assessing the Cancer Risks of Low-Level Ionizing Radiation Exposure*, Vol. 2, EMD-81-1, 2 January 1982.
3. M. Kaku and J. Trainer, 'Nuclear power: both sides', Chapter 2, *Underestimating the Risks* by K. Z. Morgan, W. W. Norton & Co., New York and London.
4. A. M. Stewart, Delayed effects of A-bomb radiation: a review of recent mortality rates and risk estimates for five-year survivors, *J. of Epidemiology and Community Health*, **36**, 80–86 (1982).

5. McClenahan, Wasted X-rays, *Radiology*, **96**, 453 (Aug. 1970).
6. A. M. Stewart and G. Kneale, Radiation dose effects in relation to obstetric X-rays and childhood cancer, *Lancet*, 1185, 6 June 1970.
7. ICRP-6, *Recommendations of ICRP*, 1962.
8. ICRP-26, *Recommendations of ICRP*, 1977.
9. G. W. Kneale, and A. M. Stewart, Prenatal X-rays and cancer: further tests of data from the Oxford Survey of Childhood Cancers, *Health Phys.*, **51**, 3, 369 (1986).
10. Nader, R., Wake up America, unsafe X-rays. *Ladies Home Journal*, **85**, 126, (1968).
11. U.S. Federal Radiation Council, *Guidance for the Control of Radiation Hazards in Uranium Mining*, Report No. 8 (revised), Sept. 1967, US Supt of Documents, US Govt Printing Office, Washington DC 20402, US.
12. ICRP-24 *Radiation Protection in Uranium and Other Mines*, 1977.
13. ICRP-16, *Protection of the Patient in X-Ray Diagnosis*, 1970.
14. ICRP-34, *Protection of the Patient in Diagnostic Radiology*, 1982.
15. US Federal Radiation Council, *Revised Fallout Estimates for 1964–1965 and Verification of the 1963 Predictions*, Report No. 6, October 1964, US Supt of Documents, US Govt Printing Office, Washington, DC 20402, US.
16. US Federal Radiation Council, *Estimates and Evaluation of Fallout in the United States from Nuclear Weapons Testing Conducted through 1962*, Report No. 4, May 1963, US Supt of Documents, US Govt Printing Office, Washington, DC 20402, US.
17. ICRP-22, *Implications of Commission Recommendations That Doses Be Kept as Low as Readily Achievable*, 1973.
18. BEIR-III Report, *The Effects on Populations of Exposure to Low Levels of Ionizing Radiation*: *1980*, by Committee on the Biological Effects of Ionizing Radiations, National Research Council.
19. H. Lisco, M. P. Finkel and A. M. Brues, 'Carcinogenic Properties of Radioactive Fission Products and Plutonium, 'Thirty-Second Annual Meeting of the Radiological Soc. of North Amer., 2 Dec. 1946.
20. ICRP-40, *Protection of the Public in the Event of Major Radiation Accidents*: *Principles for Planning*, 1984.
21. K. Z. Morgan and J. E. Turner, *Principles of Radiation Protection*, R. E. Krieger Pub. Co., 1973.
22. C. J. Johnson, Cancer incidence in an area contaminated with radionuclides near a nuclear installation, *Ambio*, **10**, 176 (1981).
23. B. Franke and R. Alvarez, External gamma radiation around the Savannah River plant, *Ambio*, **14**, 2, 104 (1983).
24. Z. Medvedev, Facts behind the Soviet nuclear disaster, *New Scientist*, 30 June 1977.
25. Z. Medvedev, Winged messengers of disaster, *New Scientist*, November 1977.
26. D. W. Moeller, L. C. Sunt and W. R. Casto, Personnel overexposures at commercial nuclear power plants, *Nuclear Safety*, **22**, 4, 498 (July–Aug 1981).
27. R. E. Kasperson, Closing the protection gap: setting health standards for nuclear power workers, *Environment*, **24**, 10, 14 (Dec 1982).
28. ICRP-30, *Limits for Intakes of Radionuclides by Workers*, Part 1, Vol 2, No. 3/4, 1979; Part 1, Vol 3, No. 1–4, 1979; Part 2, Vol. 4, No. 3/4, 1980; Part 2, Vol 5, No. 1–6, 1981; Part 3, Vol. 6, No. 2/3, 1981; Part 3, Vol. 7, No. 1–3, 1982.
29. ICRP-42, *A Compilation of the Major Concepts and Quantities in Use by ICRP*, 1984.

APPENDIX I

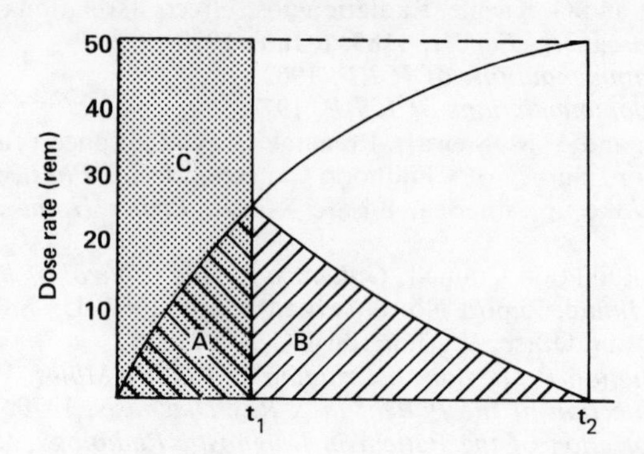

To show that area A + area B = area C.

$$R = \frac{1.87 \times 10^4 \mathrm{P}\ \overline{\mathrm{EQN}}\ (1 - \mathrm{e}^{-\lambda t})}{\lambda m} = C_1 \frac{(1 - \mathrm{e}^{-\lambda t})}{\lambda} \text{ rem/y}$$

$$A = \int_0^t R\, \mathrm{d}t = \frac{C_1}{\lambda^2} (\lambda t_1 + \mathrm{e}^{-\lambda t_1} - 1)$$

$$B = \frac{C_1(1 - \mathrm{e}^{-\lambda t_1})}{\lambda} \int_0^{t_2} \mathrm{e}^{-\lambda t}\, \mathrm{d}t = \frac{C_1(1 - \mathrm{e}^{-\lambda t_2})}{\lambda^2} (1 - \mathrm{e}^{-\lambda t_1})$$

$$= \frac{C_1}{\lambda^2} (1 - \mathrm{e}^{-\lambda t_2} - \mathrm{e}^{-\lambda t_1} + \mathrm{e}^{-\lambda t_1} \mathrm{e}^{-\lambda t_2})$$

$$A + B = \frac{C_1}{\lambda^2} (\lambda t_1 - \mathrm{e}^{-\lambda t_2} + \mathrm{e}^{-\lambda t_1} \mathrm{e}^{-\lambda t_2}) \qquad (1)$$

$$t_1 R_{(t_1 + t_2)} = \frac{C_1 [1 - \mathrm{e}^{-\lambda(t_1 + t_2)}]\, t_1}{\lambda} = \frac{C_1}{\lambda^2} (\lambda t_1 - \lambda t_1 \mathrm{e}^{-\lambda t_1} \mathrm{e}^{-\lambda t_2}) \qquad (2)$$

and

$$A + B = t_1 R_{(t_1 + t_2)} \text{ if } -\mathrm{e}^{-\lambda t_2} + \mathrm{e}^{-\lambda t_1} \mathrm{e}^{-\lambda t_2} = -\lambda t_1 \mathrm{e}^{-\lambda t_1} \mathrm{e}^{-\lambda t_2}$$

multiplying by $\mathrm{e}^{\lambda t_2}$ we have $-1 + \mathrm{e}^{-\lambda t_1} = -\lambda t_1 \mathrm{e}^{-\lambda t_1}$ and $\mathrm{e}^{\lambda t_1} = 1 + \lambda t_1$.

Expanding by Maclaurin's Series,

$$\mathrm{e}^{\lambda t_1} = 1 + \lambda t_1 + \frac{\lambda^2 t_1^2}{2} + \frac{\lambda^3 t_1^3}{6} + \frac{\lambda^4 t_1^4}{24} + \cdots$$

If $\lambda t_1 \ll 1, \mathrm{e}^{t_1} = 1 + \lambda t_1$.

For

$$\text{Sr-90 } \lambda t_1 = \frac{0.693}{\mathrm{T}} = \frac{0.693}{17.53} = 0.04 \quad \text{and} \quad 1 + \lambda t_1 = 1.04$$

while

$$\frac{\lambda^2 t_1^2}{2}\,\frac{\lambda^3 t_1^3}{6}\,\frac{\lambda^4 t^4}{24} = 3.6\times10^{-6} + 3.2\times10^{-9} + 2.1\times10^{-2} \text{ or } \ll 1.04.$$

Or for Sr-90, equation (1) above gives

$$A + B = \frac{6.9905}{(0.1397)^2} = (0.1397 - 0.001 + 0.0009) = 50 \text{ rem}$$

and from equation (2)

$$t_1 R_{(t_1 + t_2)} = \frac{6.9905}{(0.1397)^2}\,(0.1397 - 0.00013) = 49.99 \text{ rem.}$$

For I-131 (using ICRP-2 values),

$$A + B = \frac{1120}{(33.28)^2}\,(33.28 - 0 + 0) = 33.7 \text{ rem and } t_1 R_{(t_1 + t_2)}$$

$$= \frac{1120}{(33.28)^2}\,(33.28 - 0) = 33.7.$$

APPENDIX II

Equations used

$$R\left(\frac{\text{rem}}{\text{y}}\right) = qf_2(\mu\text{Ci})\, 3.7\times10^4 \left(\frac{\text{dis}}{\text{S }\mu\text{Ci}}\right) 3600\times24\times365\left(\frac{\text{S}}{\text{y}}\right)\overline{\text{EQN}}\left(\frac{\text{MeV}}{\text{dis}}\cdot\frac{\text{rem}}{\text{rad}}\right)$$

$$\times 1.602\times10^{-6}\left(\frac{\text{erg}}{\text{MeV}}\right)\frac{1}{m}\left(\frac{1}{\text{g}}\right)\frac{1}{100}\left(\frac{\text{g rad}}{\text{erg}}\right)$$

$$= 1.869\times10^4 \frac{qf_2\times\overline{\text{EQN}}}{m}$$

$$P\left(\frac{\mu\text{Ci}}{\text{y}}\right) = 6.9\times10^6 f_\text{a}\left(\frac{\text{cc}}{\text{d}}\right) 365\left(\frac{\text{d}}{\text{y}}\right)(\text{MPC})_\text{a}\left(\frac{\mu\text{Ci}}{\text{cc}}\right) = 2.518\times10^9 f_\text{a}\,(\text{MPC})_\text{a}$$

$$\frac{\text{d}q}{\text{d}t} + \lambda q = P \quad \text{or} \quad q(\mu\text{Ci}) = \frac{P}{\lambda}(1 - \text{e}^{-\lambda t})$$

$$R\left(\frac{\text{rem}}{\text{y}}\right) = \frac{(\text{MPC})_\text{a}\overline{\text{EQN}}\, f_\text{a}(1 - \text{e}^{-\lambda t})}{2.124\times10^{-14}\text{m}\lambda}$$

$$D(\text{rem}) = \int_0^t R\,\text{d}t = \frac{(\text{MPC})_\text{a}\,\overline{\text{EQN}}\, f_\text{a}}{2.124\times10^{-14}\,\text{m}\lambda^2}\times(\lambda t + \text{e}^{-\lambda t} - 1)$$

Using values from ICRP-2

$$R_{\text{for Sr-90}} = \frac{3.1 \times 10^{-10} \times 5.5 \times 0.12\ (1 - e^{-0.0395t})}{2.124 \times 10^{-14} \times 7 \times 10^{3} \times 0.0395}$$

$$= 34.8(1 - e^{0.0395t})_{\text{rem/y}} \text{ to bone}$$

$$R_{\text{for I-131}} = \frac{9 \times 10^{-9} \times 0.23 \times 0.23\ (1 - e^{-33.28t})}{2.124 \times 10^{-14} \times 20 \times 33.28}$$

$$= 33.67(1 - e^{-33.28t})_{\text{rem/y}} \text{ to thyroid}$$

$$D_{\text{for Sr-90}} = 881(0.0395t + e^{-0.0395t} - 1) \text{ rem to bone surfaces}$$

$$D_{\text{for I-131}} = 1.01(33.28t + e^{-33.28t} - 1) \text{ rem to thyroid}$$

Using values from ICRP-30

$$R_{\text{for Sr-90}} = \frac{1.62 \times 10^{-9} \times 1.1 \times 0.01\ (1 - e^{-0.140t})}{2.124 \times 10^{-14} \times 120 \times 0.140}$$

$$= 50\ (1 - e^{-0.14t})_{\text{rem/y}} \text{ to bone}$$

$$R_{\text{for I-131}} = \frac{1.9 \times 10^{-8} \times 0.23 \times 0.3\ (1 - e^{-33.73t})}{2.124 \times 10^{-14} \times 20 \times 33.73}$$

$$= 91\ (1 - e^{-33.73t})_{\text{rem/y}} \text{ to thyroid}$$

$$D_{\text{for Sr-90}} = 357(0.14t + e^{-0.14t} - 1) \text{ rem to bone surfaces}$$

$$D_{\text{for I-131}} = 2.71(33.73t + e^{-33.73t} - 1) \text{ rem to thyroid}$$

Radiation and Health
Edited by R. Russell Jones and R. Southwood

12
Discussion Period 2

Dr Lesna (*Pathologist, Royal Victoria Hospital, Bournemouth*) Professor Berry, you showed data on cancer fatalities amongst radiologists up to 1949. My husband has been a radiologist for the past 20 years. He has been exposed and measured and nobody seems to follow it up. What is the permitted dose for radiologists?

Professor Berry 1 milliSievert (100 mrem) per annum is the dose limit for members of the general public. For occupationally exposed groups such as radiologists the dose limit is 50 mSv (5 rems) per annum in the UK, but doses above 15 mSv (1.5 rem) would require investigation. Now it may well be that health authorities can decide to use the new ionizing regulations, as a means of reducing further the radiation doses to those who work in hospitals.

Indeed, the health authority for which I work, Bloomsbury, accepted a recommendation of the Radiation Safety Advisory Committee which I chair, that we will institute radiologically safe working procedures, so that the staff working in diagnostic nuclear medicine will not even classify as radiation workers. That means keeping exposures below 5 mSv (500 mrem) per annum, one tenth of the occupational dose limit.

Mr Tony Webb (*Radiation and Health Information Service*) I wonder if you can help me to understand something that I have never understood about the concept of ALARA. You stress the importance of keeping doses as low as reasonably achievable; yet ICRP 30 lays down a set of limits for internal organ exposure which are between two and ten times higher than the previous ICRP limits. At the same time ICRP members reassure us that under no circumstances should these new dose limits be used to permit higher exposures where the old system was able to achieve lower ones. You have a clear case where they were reasonably achievable before, why are they not achievable under the new regulations? And why is there no guidance to national governments as to which set of limits they should be using?

Professor Berry The organ dose in the past were designed to prevent non-stochastic damage, and if you observe the dose limits at present, the dose to any one organ will not exceed the minimum dose at which you will actually produce non-stochastic damage. Certainly we will be looking in our review next year, at whether this has caused confusion in national regulations, and if so, we may well go back to an organ-based system. But certainly, the response we have had so far, indicates that ALARA is keeping doses down.

Mr Webb I understand that. But if I can give you just three brief examples. For workers exposed to radioiodines in isotope production, or in nuclear medicine, their intakes from radioiodines are permitted to rise by between 120 and 400 per cent. Similarly, workers in the Candu plants in Canada receive significant exposure from tritium, and we have already heard from Dr Beral about the possible hazards from tritum. Yet the new ICRP system permits a 34 per cent increase in tritium exposure.

Then there are the uranium miners, many of whom are receiving the bulk of their dose from radon exposures. Yet under the new ICRP regulations the radon lung dose is permitted to rise by nearly 40 per cent, even though there is an epidemic of lung cancers amongst uranium miners and the epidemiological data is showing higher lung cancer rates than expected. Those are three cases where the ICRP has patently failed to offer practical advice in terms of keeping doses as low as reasonably achievable. I think it reinforces the case that David Gee made, that the concept is fine, but in practice it does not work!

Professor Berry I hear that and we will be looking at it next year as the Commission, to see if it really is not working, but I must come back to your point about the uranium miners. A major influence on the epidemic of lung cancer among uranium miners, is smoking. So it is not only radiation. It may be of course that the two are synergistic, so that you have a greatly increased hazard in that particular circumstance. The point again, is that the dose limit sets the range of unacceptability, and the fundamental aim is to get doses as low as reasonably achievable.

Dr Russell Jones I think there are two quite separate issues to which we should perhaps address ourselves. One is quite simple: has the ICRP got the science right? Is the 125 cancer deaths per million person rem an accurate reflection of the scientific data? The second question is perhaps more complicated and has considerable social and political overtones. Are the members of the ICRP in fact the right people to be saying what an acceptable or unacceptable limit is? On the first point I would like to pick up one aspect of Professor Radford's presentation. To my knowledge the data that you presented, pooling the Hiroshima and Nagasaki survivor data, is in fact the most up-to-date data that is currently available.

Professor Radford Yes.

Dr Russell Jones And you gave some figures for the low dose end of your curve which was suggesting that the doubling dose for all cancers, excluding leukaemia, was in the range of 107 to 175 rem. Is that right?

Professor Radford That is correct.

Dr Russell Jones Can you tell us how that compares with the figures that the ICRP gives?

Professor Radford Well, the problem in going from doubling doses to absolute risk is that it becomes a function of the background cancer rates that exist in the population. Even so, if you assume that 15 per cent of deaths result from cancer, that is 150 000 per million deaths, then 125 deaths per million per rem represent one out of 1200 per rem which implies a doubling dose of 1200 rem. So we are talking about a difference of ten from ICRP. And, if the risks for the younger segment of the population end up producing a higher risk for the whole population, that would lower it even further. If the lower end of the curve is reliable, it suggests that the doubling dose for all cancers, including leukaemia, is about 130 rem.

Dr Russell Jones There has been a lot of criticism of the bomb data and some commentators have tried to dismiss the bomb data on the basis that it may be rather unreliable. Could I ask what your view of the updated bomb data is, in comparison with other data that we have on the human population?

Professor Radford Sarah Darby did a comparison while she was in Hiroshima, comparing the ankylosing spondylitis follow-up study that Sir Richard Doll began, and found that the two studies agreed remarkably well, taking into account the new dosimetry. So the Japanese data seem remarkably consistent with that study. If you dismiss the bomb data you have to dismiss a lot of other studies as well. Nevertheless there are still some unresolved issues. For example, are the risks of radiation independent of the base line of cancer incidence in different populations? In Japan, for instance, the breast cancer rate among women is significantly lower than it is in Western women. And yet the absolute risk for radiation-induced breast cancer is very similar between Japanese and Western countries. Second, the children who were *in utero* at the time of the atomic explosions, why do they not demonstrate an excess of leukaemia in the ten years following the bombing? They are showing an excess risk of cancer now, but that is quite different from what has been observed in the British children who are irradiated *in utero*. So there are discrepancies. But on the whole I would say that other studies have tended to support the Japanese data.

Dr R. Belbeoch (*GSIEN Paris—The Group of Scientists for Information on Nuclear Energy*) It is important to realize that the atomic bomb survivors were not a normal population, for two reasons. First, only the healthiest were surviving—and second, the immunosuppressive effects of radiation means that many survivors with increased cancer risk died from infection before the cancer had been expressed.

I think that the Chernobyl population which was evacuated, the 135 000 people whose doses range from 5 rad to 54 rad, will probably prove to be a better representation of the normal population.

Professor Radford I have read an enormous literature on the conditions that existed in Japan both before and after the bombing. There is no question that there was deprivation, and there was a lot of malnutrition. To say that they were a perfectly normal healthy population is I think wrong. On the other hand, the assumption is made that these conditions will somehow affect cancer risk thirty, forty or fifty years later. Now that is an assumption for which we have little evidence. But if that were so, then the risks derived from the Japanese data would differ markedly from the risks derived from other studies. And as I just said, they do not.

Professor Berry I would like to comment on the Chernobyl situation. One of the things one hopes to achieve from this terrible tragedy is that we will obtain reliable epidemiological data. Unfortunately it is going to take the next forty or fifty years to get it. Also for the single large population that was evacuated from the 30 kilometre circle around Chernobyl, the problem will be to find a suitable control population. All of the other populations that have been studied—particularly worker populations—are out because they are selected because they are healthy. Ankylosing spondylitics are out because they are sick. You take your choice. The point that Professor Radford makes is a good one; that from several independent sources we are coming to relatively consistent estimates in the rate of cancer induction as a function of radiation dose. Now, whether those final figures are a factor of 2, or a factor of 5 different from the current ICRP estimates remains to be determined.

Dr Russell Jones A figure of 125 cancer deaths per million person RAD was presumably the best figure in 1977 as far as your committee was concerned?

Professor Berry Yes.

Dr Russell Jones Do you think that will still be the best figure or do you think that will change?

Professor Berry Well, I was not a member of the ICRP, so I will not take responsibility for 1977. Obviously we are taking account of all the new data and the situation is being kept under constant review.

Mr Frank Cook, MP We have heard so much this afternoon about different acceptable limits. We have heard about the 5 mSv (500 millirem) statutory limit for members of the public and the 1 mSv (100 rem) recommended limitation which was introduced in February this year. On television, I made the point that the limitations here are somewhat less safe than those that apply in the United States, or in Western Germany. The Minister for the Environment, Mr Waldegrave, took issue with this and accused me of being alarmist. Since the NRPB is represented here today, could someone advise me whether in fact the UK limitations, either recommended or statutory, are safer, less safe, or equal to the standards that apply in the United States and in West Germany?

Professor Radford I get the uncomfortable feeling that the reason why the 100 millirem was adopted by the NRPB, was that British Nuclear Fuels told them they could achieve it. I am sorry to have to say that the practices in the industry have been riding the standards for a long, long time. And I think that is a very sad commentary on the independence of these regulatory groups. I would just add one final note, since we have an ICRP member on the panel. I do not think that scientists are particularly good, or appropriate, to be making moral and social judgements on how much radiation exposure is acceptable to the public.

Professor Berry I am going to support Professor Radford on this point. I do not think that the scientist, by virtue of his science, has any right to believe that he is better at making moral and social judgements. By presenting knowledge in a clear and understandable way, he can, however, assist the moral decision.

Mr Cook Are the standards of the UK less safe, more safe, or equal to, that was my question. Will someone answer it please?

Dr Russell Jones I think that Dr Roger Clarke could perhaps give you a direct answer as secretary of the National Radiological Protection Board.

Dr Clarke I think it is a difference in the application of ALARA. As I understand it, the UK and the USA have the same primary dose limits for members of the public—let us say 5 mSv or 500 mrem per year. In America, the EPA set a uranium fuel cycle standard which, on the basis of what they thought was as low as reasonably achievable, led to a dose specification of 25 millirem, rather than 500 mrem, which they set for operating nuclear reactors. It does not apply to other sources. So in this country what government has done is to start with a dose limit and look at authorizations. The Department of the Environment and MAFF work on a case by case basis. In general, the offsite doses at most of our operating reactors, would probably not be very different from the sort of figures in the US EPA uranium fuel cycle standards. The one exception of course is Sellafield.

Dr Russell Jones Could I just ask you what the legal position of the 100 millirems is?

Dr Clarke The major legal instrument in this country is the Health and Safety Executive's ionizing radiation regulations which put a dose limit for members of the public of 500 mrem to any individual. Certainly we have advised that it would be prudent for long-term planning not to exceed 100 mrem, and because people are exposed to various sources, the 100 mrem should not all derive from a single source.

Mr Cook Could I just add one comment? If it is reasonably achievable for the USA to make a 25 mrem a reasonably achievable target and 30 mrem in Western Germany, it ought to be reasonably achievable in the UK.

Dr David Sowby (*formerly Secretary of ICRP*) From what I have heard this afternoon, it sounds as if I have been employed by a body that would like to make the Mafia a choir of angels. There was no domination of the ICRP by physicists, there is no domination by physicists. But to make a more practical point, there is too much concentration on the question of dose limits. Roger Berry has stated that the dose limits represent the point at which you enter the area of unacceptability. You should never reach those limits, except occasionally. As for a continuous exposure, at what Dr Morgan calls the maximum permissible concentrations, that is a term that has gone out now, and would represent an utterly intolerable situation. The emphasis now, is on keeping all doses as low as reasonably achievable.

Dr Russell Jones Presumably though, if the employer did go up to the limit, he would not be doing anything illegal? It would still be legally permissible. Is that right?

Dr Sowby Yes, as long as he is below or up to the limit. In most countries that is still legal. But you would want to know why he was up to the limit and if it continued you would certainly want to know why. And the Commission says so in Publication 26 I may say.

Mr David Gee The unions do understand the philosophy of ALARA, but it is nothing new. When the ICRP introduced ALARA in 1977, the rest of occupational health was already using 'as low as reasonably practicable'. They just called it ALARP instead of ALARA. But all the practitioners in this field, including the Enforcement Inspectorate, the employers, unions and others, recognize that ALARP is more or less a fiction. It is impossible to properly quantify it by situation, by limit, by time, by whatever. It is so 'fog-like' as the inspectors put it, that it is basically unusable. All they can enforce are the numbers and that is what the real world is about. There is no real pressure to reduce unnecessarily dirty practices unless you start by bringing down the limits.

Professor Berry I hear what you say and I think ICRP will be looking at this again next year. In this country, however, not ICRP but nationally, HSE has interpreted the ICRP recommendations by imposing an investigation level. If you intend to exceed one-third of the dose limit then you have to satisfy an inspector. But I hear what you say and I think it is one of the things we are going to have to look at again, to see if it really does work and if not, then we will have to go back to the system of rigid dose limits. Now the other approach is to impose financial penalties so that the closer you get to the dose limit, the more it costs. Similarly, the closer you get to the lower end, the less it costs. So you have at least a concept that people can all understand. You rightly point to how difficult it has been changing over to SI units. None of us actually have a feel yet for the size of the new units, and the fact that the Russians do not use them at all does not help. But the point about cost is that it relates to units which we use in our everyday lives.

Dr Russell Jones Could I just ask on a point of information Professor Berry, when you consider setting dose limits, does the ICRP ask for an actual economic analysis of what that would cost the industry to implement?

Professor Berry I cannot speak for anything before last year. Certainly it is one of the things we have to consider, in making recommendations.

Dr Russell Jones Professor Morgan, in your experience does cost carry a large amount of weight in the deliberations of ICRP?

Professor Morgan Yes, I think it does. As I indicated our nuclear regulatory commission set a value of $1000 per person rem and it cost $1000 for industry to make this correction. But it seems to me that there is far too wide a margin between the maximum permissible dose and what you expect to get on the basis of ALARA. It is like wanting a speed limit of 55 mph, but setting it at 500 mph because most people drive at 50 mph anyway. Consequently there are some fairly sizeable groups that are getting a large fraction of the maximum permissible dose both internal and external, and I think that is unacceptable. There are quite a few people in industry receiving close to 30 rems a year from exposure to strontium, that is integrated over a 50-year period, and there are others that are getting close to 30 rems a year from iodine. And you have to multiply that by about 50 to calculate the total dose over a 50-year period. And that of course is completely unacceptable because it means that you may have a more than 50 per cent risk of developing thyroid carcinoma.

Ms Jean Emery (*Greenpeace*) Peter Taylor showed us that if you live in Cumbria, as I do, and if you eat local fish and shellfish, and you are a member of the critical group, then in 1981 you would have received 350 mrem, the next year 250 mrem. These figures do not include doses from previous years, the doses from the 1957 fire, the fact that we are a high rainfall area, or the effects of Chernobyl in Cumbria. Given that many of us will have exceeded the 100 mrem limit for several years, can Dr Berry advise us on how to keep our exposures below 100 mrem, because neither the ICRP nor the NRPB are providing the radiological advice and protection which are needed. In fact the only reason that people have stopped eating local shellfish is because of the actions of pressure groups. Frankly, that is not good enough. People are being paid to do a job and we want more practical advice as to how to avoid these higher doses.

Professor Berry As you say, the pressure has been inexorably downward to reduce discharges from the Sellafield plant, so that members of the critical group who derive most of their protein from shellfish caught locally, will not exceed the ICRP dose limit over a lifetime. Yes, things were bad a few years ago, I think there is no question of that. I think it is the pressure from the independent public bodies like the NRPB which is helping to reduce the discharges.

Ms Emery 100 mrem is based, if you live long enough, on a 70-year life-span. How can I achieve that average dose, when I have already received doses well above the average?

Dr Russell Jones I think the problem is that in Sellafield the doses to the public were approaching 500 millirems in the 1970s, and even now they are over 100 millirems, so how does this square with the 100 millirems recommendation?

Professor Berry I think there is a real problem. The basic problem is that you are committed to a certain dose as a result of previous discharges, and the simpler answer is to say 'OK if you don't liked it, move away'. But that is not an acceptable answer. The answer is to reduce the contribution to your radiation exposure from those sources which you can control. Radiation from the sun you cannot control. The contribution from weapons' fallout you cannot control because it happened 20 years ago. So you deal with those sources which you can control. And certainly the pressure from the public bodies, has been to progressively reduce the additional man-made radiation.

Professor Wolfgang Jacobi I am from Germany and I am a member of the ICRP and I want to comment on the 30 mrem per annum limit which was mentioned for the Federal Republic of Germany for nuclear power plants. The history of this limit is that it is an authorized limit based on the practical experience of nuclear power stations. During the development of these stations and following technical improvements, it was shown that such a level could be reached and therefore our regulations re-express this limit as an authorized limit. But this is not a limit in the sense of ICRP limits, which mark the lower level of an unacceptable region. With reference to the ICRP occupational limit, I think the main thing is what risk is acceptable to a worker. Of course, this problem is not unique to radiation workers; it is a problem for all types of industry and I am not sure that any industry has resolved this satisfactorily. ICRP proposed a level of exposure which produced a risk of fatality comparable to the risks of dying in other industries. Please note that ICRP can only recommend, and I think it would be a very good idea for ICRP if some other international organization would advise on what should be an acceptable risk.

Dr Russell Jones That is one of the reasons why we organized the conference.

Mr Gee If the current limit is five times too high, it may be more, as we heard from Professor Radford; but assuming it is only five times too high, then that will result in 10–15 per cent of radiation workers who will die over their lifetime from radiation-induced cancer, if exposed to the limit. And that limit is therefore beyond all bounds of acceptability. If the intention is to fix all limits on the basis of acceptability then there is a clear case for an immediate fivefold reduction in that limit. At the present time the ICRP does not receive any input from organizations that represent those people who actually run the risks associated with these limits. So why are radiation workers not represented on the ICRP? Or failing that, why do the ICRP not accept a delegation from an internationally based organization. I speak not just as a member of GMB, the General, Municipal, Boilermakers and Allied Trades Union, but also as a representative of ICEF. ICEF represents atomic power workers and chemical workers in dozens of countries throughout the world. They too hold the view, expressed most recently at their Toronto conference only three weeks ago, that the ICRP dose limits are too high by a factor of five. And so we would like to come to the next ICRP meeting in Como, if not before, and argue the case for an immediate fivefold reduction. I hope that you will feel able to respond to this request.

Second, my organization does not accept the philosophy of your index of harm approach. We dislike comparisons with other industries as a yardstick for action. If one considers trawlermen, coalminers, fishmongers and shopkeepers, then clearly the trawlermen are involved in the most hazardous industry. But that does not mean that safety measures can be ignored in the other occupations. So your fundamental approach which tries to identify an acceptable level of death or detriment in society, and then fix a radiation dose to produce an equivalent effect is basically a nonsense.

Part 4

Epidemiological Data

Radiation and Health
Edited by R. Russell Jones and R. Southwood

13

Variation in Individual Sensitivity to Ionizing Radiation

PAUL D. LEWIS
Royal Postgraduate Medical School, Hammersmith Hospital, London, UK

Individuals vary widely in the susceptibility of their body tissues to the damaging effects of ionizing radiation. Current knowledge of individual variation is fragmentary; the problem has not been studied in depth and laboratory methods for assessing radiosensitivity are at present too laborious for large populations to be examined. Much of the relatively little that is known in this field comes from research on patients with rare hereditary diseases, several of which show striking involvement of the nervous system. Nevertheless, it has become apparent that variation in individual susceptibility to ionizing radiation is an issue of importance for the population at large. Very many cancer patients receive radiotherapy, and a proportion of these are found to be so hypersensitive to their treatment, in terms of the response to radiation of their normal tissues, that therapy has to be modified or stopped prematurely. Available evidence suggests that these sensitive patients are representative of the general population. The possibility that some 'normal' people who may be occupationally exposed to ionizing radiation are comparably hypersensitive has to be considered in the formulation of radiation protection advice. Both increased susceptibility to radiation and unusual radio-resistance may be associated with cancer-proneness. Marked resistance to the damaging effects of radiation might be a factor in cases of unexpected, and apparently inexplicable survival after radiation exposure at lethal levels.

ATAXIA TELANGIECTASIA

The first disease in which marked hypersensitivity of body tissues to radiation was shown was ataxia telangiectasia (AT). This rare disorder of infancy and childhood was first recognized in the inter-war years,[1,2] but did not become widely known until the late 1950s.[3,4] It is a recessively inherited disease, sufferers from which develop progressive ataxic disability due to degenerative changes in cerebellum, brainstem and spinal cord; recurrent lung infections due to innate immunodeficiency; and dilated blood vessels (telangiectases) of the conjunctiva and upper face. Malignant tumours, generally of lymphoid tissues or blood but occasionally arising from the alimentary tract or elsewhere, occur in 10 per cent of cases.[5] Patients with this disease usually die in childhood as a result of chest infection, but in the minority the cause of death is malignant disease. Survival into adult life may occur in mild cases, though this is uncommon.[6]

It was misadventure in the course of treatment of cancer that led to the discovery of extreme hypersensitivity to radiation of normal cells and tissues in AT. A ten-year-old boy with the disease developed a lymphosarcoma in his palate, and radiotherapy was given[7]. The plan of treatment was that a tumour dose of 4000 rads should be delivered. However, after receiving 3000 rads in fifteen doses over a three-week period his skin and oral mucosa showed a severe reaction. Pain, swelling and ulceration prevented his eating, and eventually gastrostomy had to be performed. He died nine months after radiation. Autopsy showed healed radiation dermatitis and skin ulceration in the treated zone, and massive ulceration of pharynx and tongue, apparently also due to radiation rather than to residual or recurrent tumour. A year later in 1968, a second and similar case was reported in the same medical journal.[8] This nine-year-old boy was found to have a growth in the left side of his chest, and lymph node biopsy showed Hodgkin's disease. A planned course of radiotheraphy of 4000 rads to the chest was begun, but after 2750 rads over two weeks severe difficulty in swallowing was experienced. This was due to oesophagitis. A severe skin reaction also appeared in the area that had received radiation. The patient died three months later. At autopsy no oesophageal ulceration was seen. A third case of this type was the one that initiated the relevant laboratory investigation.[9] Here a seven-year-old boy with AT and a malignant lymphoma in the right side of the chest developed skin reddening and difficulty in swallowing after 2000 rads over two weeks, and treatment had to be stopped short after 3000 rads. Skin sloughed from the irradiated areas of his chest, and he died a few weeks after treatment.

Several years earlier, an important discovery had been made about the inherited disease xeroderma pigmentosum (XP), a disorder characterized by an extreme proneness to sunburn, with scarring of light-exposed skin and the development of skin cancers. Skin cells from patients with XP were found to be deficient in their capacity to repair ultraviolet induced damage to DNA,[10] and when cultured in the laboratory they were shown to be far more sensitive than normal cells to the growth-inhibiting effect of ultraviolet light. In 1975, prompted by this work and the third case described above, Taylor and colleagues examined the sensitivity to X-irradiation of cultured fibroblasts from three patients with AT.[11] When they were exposed to increasing doses of radiation under optimal growth conditions in culture, such cells showed a progressive reduction in their ability to form colonies that was far greater than with control cells. Taylor *et al.* found that with normal cells 1 per cent survived to form colonies after a dose of 600–750 rads. The three AT cell strains were reduced to 1 per cent survival with doses of only 200–300 rads. The finding of marked cellular hypersensitivity to radiation in AT clearly explained the clinical observations already noted, and has since been confirmed in many other AT cases.[12] Disease of the nervous system is a major feature of AT and may also be prominent in XP.[13] Andrews and colleagues suggested[14,15] that in these DNA-repair-deficient disorders, failure of maintenance of the integrity of nerve cell DNA in the face of chronic damage was responsible for the neurological disease. It was further proposed that in many nervous systems diseases of unknown cause ('degenerative diseases') defective DNA repair might be a primary pathogenetic mechanism. Laboratory studies by Robbins and his colleagues have since produced data supporting this proposal. In a wide range of neurological disorders, including such relatively common conditions as Alzheimer's disease,[16] Parkinson's disease[16] and Duchenne muscular dystrophy,[17] as well as rare ones like familial dysautonomia,[18] tuberous sclerosis [19] and Usher's syndrome,[20] these workers have found increased cellular sensitivity to ionizing radiation. Other laboratories have not always been able to confirm these findings, which are based on techniques which are not the most widely used in the study of radiation sensitivity, and the hypothesis of Andrews *et al.* is now seen to lack support; if there is a connection between faulty DNA repair and neurological disease, it is not a causal one (see below). Only in a few diseases of the nervous system is there unequivocal evidence of increased sensitivity to radiation, implying defective DNA reparative mechanisms. These include a particular type of retinoblastoma,[21] Down's syndrome[22] and Friedreich's ataxia, the last of which will be discussed more fully. Retinoblastoma cells are less sensitive than AT cells but more sensitive than controls. Down's syndrome cells show increased chromosomal damage after radiation and decreased survival in culture experiments. Some families with Huntington's disease may also be radiosensitive,[23] but others are not. All these findings indicate that in the population at large there are individuals who probably for genetic reasons are predisposed to the damaging cellular effects of ionizing radiation.

FRIEDREICH'S ATAXIA

In this uncommon inherited neurological disorder (due to a recessive gene with a heterozygote frequency of 1 in 110)[24] increased cellular radiosensitivity was first shown in 1979 in a study of skin fibroblasts from a severely affected 21-year-old man with typical signs of the disease.[25] Friedreich's ataxia shares certain clinical and pathological features with both XP and AT, and tests of radiosensitivity in Friedreich's ataxia (FA) were initiated partly as a result of the speculations of Robbins and colleagues, and partly on the supposition that some chronic occult AT cases might be revealed. In both the original FA case and in a larger series published three years later[26] the degree of radiation hypersensitivity was found to be mild in comparison with AT but was significant when compared with findings from a control population composed of disease-free individuals, as well as cells from patients with the common sporadic neurological disorder motor neurone disease (Table 1). Fibroblasts from small groups of patients with two disorders previously claimed to show radiosensitivity (familial dysautonomia and Duchenne muscular dystrophy) and a disease said to have clinical links with FA (Charcot-Marie-Tooth disease) were also tested against controls, and in contrast to FA were found to give values for radiosensitivity within the control range (Table 2).[27]

Further evidence for the radiosensitivity of cells in FA came from studies on the repair of potentially lethal damage induced by X-rays, which was found to be impaired,[26] from analysis of DNA structure after irradiation,[28] and from examination of chromosomal aberrations caused by X-rays. In the last of these

Table 1 Effects of X-irradiation on cultured skin fibroblasts from patients with Friedrich's ataxia (FA), motor neurone disease (MND) and from control subjects

	D_0 (rads)		
	Controls	FA	MND
	132 (3)	122 (3)	150 (3)
	152 (3)	115 (2)	153 (3)
	145 (3)	104 (3)	155 (3)
	168 (3)	133 (9)	145 (3)
	145 (3)	122 (2)	150 (3)
	185 (2)	135 (2)	144 (2)
	125 (2)	120 (1)	144 (2)
	128 (2)		137 (1)
	165 (2)		
	158 (2)		
Overall mean ±SD	150.3 ± 19.2	123.7 ± 20.7[a]	147.1 ± 5.9[b]

In these experiments, cells were irradiated at graded doses, and after 14–20 days growth the colonies derived from surviving cells were counted. Dose–survival curves were constructed and the D_0 values—the dose in rads which reduces survival from any point on the exponential part of the curve to 37 per cent of the value at that point—were calculated. The lower the D_0 value, the greater the sensitivity. Experimental results are given for each case. Figures indicate mean D_0 value for each subject, with number of replicate experiments in parentheses.

[a] Significant difference from controls: $p < 0.01$.

[b] Non-significant difference from controls.

Table 2 Radiosensitivity of fibroblasts in other hereditary diseases affecting the nervous system and muscle

Familial dysautonomia		
Case	Age (Yr)	D_0 (rads)
GM0732	15	148 + 17(6)
FM2341	17	171 + 0.3(2)
GM2342	19	141 + 3.5(3)
GM2343	24	146 + 10(8)
Overall mean		151 + 6.7(19)

Charcot-Marie-Tooth disease		
Case	Age (Yr)	D_0 (rads)
Ho	58	138 ± 6.0(13)
Hu	35	144 ± 12(4)
Wal	15	139 ± 15(3)
War	57	148 ± 10(8)
Overall mean		144 ± 4.0(28)

Duchenne muscular dystrophy		
Case	Age (Yr)	D_0 (rads)
A	9	123 ± 2.9(7)
SW	4	147 ± 1.4(3)
M	?	124 ± 6.0(3)
SM	5	135 ± 7.4(4)
W	7	154 ± 4.0(3)
R	14	137 ± 8.6(5)
J	6	162 ± 1.4(3)
K	5	140 ± 3.2(4)
Total		141 ± 4.0(32)

Controls		
Case	Age (Yr)	D_0 (rads)
H	1	124 + 4.0(3)
D	11	109 + 1.5(4)
R	15	192 + 4.2(5)
B	30	159 + 9.0(12)
G	39	121 + 2.1(12)
P	41	160 + 4.3(5)
Total		144 + 12(41)

Details as in Table 1.

studies,[29] cells from groups of patients and controls were tested to measure the incidence of spontaneous abnormalities of chromosomes, and were exposed to a graded series of X-ray doses to quantify their response to radiation-induced chromosomal damage. The incidence of spontaneous aberrations in controls was at a level found in other control studies, but the background level of aberrations in FA cells was significantly raised. This finding, suggestive of a whole-body dose of 25 rads (which had not been received by any of the patients), was interpreted as indicating a hypersensitive response to routine low levels of diagnostic radiation. In culture, FA cells produced 60 per cent more aberrations than controls up to a dose of 200 rads, confirming their increased sensitivity to ionizing radiation.

The differences in measures of radiosensitivity between FA cells, on the one hand, and cells from controls and other neurological disease examined in this laboratory on the other, suggested that FA was unequivocally a hypersensitive condition and that it was probably unrelated to AT. It was felt that the neurological aspects of FA were not a direct consequence of defective DNA repair mechanisms, presumed to be responsible for the increased sensitivity to the damaging effects of radiation, as postulated by Andrews and colleagues. Rather, it was considered that FA patients possessed an abnormal gene that might be close to a chromosomal region responsible for the coding of one of the many DNA repair enzymes. Though FA patients may be genetically heterogeneous[30] as a group they show moderate relative hypersensitivity to radiation. Some normal, disease-free individuals are also relatively sensitive, while others are relatively resistant to the effects of ionizing radiation, and there is some evidence that these differences might also have a genetic basis.

VARIATIONS IN RADIOSENSITIVITY BETWEEN INDIVIDUALS IN THE GENERAL POPULATION

Laboratory studies on cells from individuals without specific medical complaints show a wide range in sensitivity to the effects of radiation. The findings on normal cells in this laboratory show a range of D_0 values (see Tables 1 and 2) from about 110 to 190 rads. Cells from several neurological diseases give values within this range, and should therefore be regarded as having normal sensitivity. Comparable figures, with a wide difference between the most and least sensitive cell strains, have been found by other workers. In their review of data obtained from a large number of different cell cultures, including some patients with AT, Arlett and Harcourt[31] found a normal range in 29 cell strains of 96 to 180 rads. Differences between a relatively sensitive strain (mean D_0 from 14 experiments, 124 rads) and a relatively radioresistant one (15 experiments, 160 rads) taken from normal individuals selected at random were statistically significant. In a companion study, Weichselbaum and colleagues[32] reported a normal range of 128 to 164 rads, based on findings in six subjects.

Cox and his colleagues[33] reported a normal range comparable with all of these (100 to 160 rads, 12 samples) and noted, as did Arlett and Harcourt, that the range showed a skewed distribution, with more representatives at the lower (sensitive) than upper (resistant) end of the range. In discussing the extreme radiosensitivity of most AT cells (D_0 often as low as 50–60 rads) and the moderate sensitivity of a few AT strains, Cox *et al.* made the point that X-ray sensitivity is a gross expression of cell biology and is probably influenced by a number of genes. It follows that a spectrum of different radiosensitivities is to be expected in the population at large; inevitably there would be overlap in sensitivity between different subpopulations, producing a continuum of values. This seems a plausible explanation of the range of findings in normals, and the grouping of FA sensitivities at the lower end of the spectrum.

That radioresistance as well as radiosensitivity has a genetic basis is suggested by the observations of Bech-Hansen and colleagues,[34] who studied eight members of a family in which a diversity of cancers (two developing after radiotherapy) had occurred over five generations. Five of the eight subjects were radioresistant compared with controls, their cultured fibroblasts giving D_{10} values up to 25 per cent above the normal range obtained from five cell strains tested in parallel. The findings raised the possibility that radioresistance in this family might be an expression of an aberrant DNA-reparative process or replicative cell cycle perturbation that predisposed to cancer.

Findings in normal individuals thus show a wide range of values for radiosensitivity, the differences possibly having a genetic basis. A twofold difference between lowest and highest D_0 values does not imply a twofold difference in sensitivity between most and least susceptible subjects. Extrapolated to lethal doses of radiation, these figures indicate a *tenfold* difference in the capacity of cells to survive. A study of cellular sensitivity to radiation in cases of anomalous survival after atomic bomb explosions might reveal a high proportion of D_0 values at the upper end of the normal range. Conversely, it is interesting to speculate on the radiosensitivity of those most severely affected in limited nuclear mishaps affecting small numbers of people. This speculation raises the issue, the further discussion of which is inappropriate here, of safe radiation limits for those who may be occupationally exposed.

VARIATIONS IN RADIOSENSITIVITY IN CANCER PATIENTS RECEIVING RADIOTHERAPY

Individual variation in the susceptibility of cancer patients to damage to normal tissues in the course of radiation treatment is a well-recognized problem. The most frequent form of damage observed in clinical practice is a skin reaction; radiation burns may be severe enough to limit treatment, as happened in the AT cases described earlier. Another important acute reaction is radiation pneumonitis, sometimes encountered with irradiation of the chest.

These reactions occur in perhaps 5 to 10 per cent of cancer patients receiving radiotherapy, and are the consequence of planned 'average' doses of radiation being excessive for the minority of hypersensitive patients. The role of individual hypersensitivity to radiation in the genesis of such important long-term consequences of radiotherapy as radiation myelitis is totally unexplored.

One patient with radiation pneumonitis after chest irradiation was found in this laboratory to have a skin fibroblast D_0 value in the normal range. However, another patient was shown to be hypersensitive on laboratory testing.[35] This 62-year-old man had a malignant lymphoma of his skin, and received 1600 rads of superficial X-rays in four fractions over five days to localized skin tumours, followed after a week by 600 rads to his whole body. A severe skin reaction soon occurred in the superficially treated areas, and he developed oesophagitis. This was of an intensity occurring usually only after a dose of 6000 rads in the treatment of oesophageal cancer; here the dose of the oesophagus was unlikely to have been more than 500 rads from his superficial treatment. In the following week a severe generalized skin reaction developed. This healed over three weeks. A subsequent whole body dose of 200 rads produced a further severe skin reaction. This eventually healed. His total treatment was less than half of that which is normally tolerated, and it produced major effects in his normal tissues. The radiosensitivity of his skin fibroblasts was tested and a D_0 value of 77 rads, comparable with that found in some patients with AT, was obtained.

Observations in this laboratory that not all radiation-sensitive radiotherapy patients show increased cellular sensitivity in laboratory tests are in accord with earlier findings of Weichselbaum and colleagues.[36] Two of the three clinically sensitive patients they studied had fibroblast D_0 values of 101 and 108 rads, below their subsequently established normal range but interpreted then as within normal limits. Such tests are slow, laborious and expensive to carry out, and there is the need for easy and rapid tests of cellular sensitivity so that individuals who may be exposed to radiation can be screened in advance of exposure. Rapid tests using lymphocytes, the radiosensitivity of which in colony-forming experiments is comparable with that of fibroblasts,[37] have been developed by Harris and colleagues in the investigation of rheumatic disease.[38] A range of findings has been shown in population studies on the elderly;[39] however, no correlations with fibroblast D_0 values have been made and these tests have yet to be validated. It is to be hoped that more research resources can be applied to the issue of determining individual radiosensitivity. Gains in radiotherapeutic practice and in the identification of those at risk to occupational or accidental radiation exposure would follow.

REFERENCES

1. A. Syllaba and K. Henner, Contribution à l'indépendence de l'athétose double

idiopathique et congénitale. Atteinte familiale, syndrome dystrophique, signe du réseau vasculaire conjonctival, intégrité psychique, *Rev. neurol*, 1926; 1, 541–62. Cited by R. P. Sedgwick and E. Boder, Ataxia-telangiectasia, in P. J. Vinken, G. W. Bruyn (eds), *Handbook of Clinical Neurology*, vol. 14 The phakomatoses. Amsterdam: North-Holland, 1972, pp 267–339.

2. D. Louis-Bar, Sur un syndrome progressif comprenant des telangiectasies capillaires cutanées et conjonctival symétriques, à disposition naevoide et des troubles cerebelleux, *Confin. neurol*, **4**, 32–42 (1941).
3. C. E. Wells and G. M. Shy, Progressive familial choreoathetosis with cutaneous telangiectasia, *J. Neurol. Neurosurg. Psychiat.*, 1957; **20**, 98–104.
4. E. Boder, and R. P. Sedgwick, Ataxia-telangiectasia. A review of 101 cases, G. Walsh. (ed.), *Little Club Clinics in Developmental Medicine, 8*. London: Heinemann, 1963, pp. 110–18.
5. B. D. Spector, A. H. Filipovich, G. S. Perry III and J. H. Kersey, Epidemiology of cancer in ataxia-telangiectasia, in B. A. Bridges and D. G. Harnden (eds), *Ataxia-telangiectasia. A Cellular and Molecular Link between Cancer, Neuropathology, and Immune Deficiency*. Chichester: Wiley, 1982, pp. 103–138.
6. R. P. Sedgwick, Neurological abnormalities in ataxia-telangiectasia, in B. A. Bridges and D. G. Harnden (eds), *Ataxia-telangiectasia. A Cellular and Molecular Link between Cancer, Neuropathology, and Immune Deficiency.* Chichester: Wiley, 1982, pp. 23–35.
7. S. P. Gotoff, E. Amirmoki and E. J. Liebner, Ataxia-telangiectasia. Neoplasia, untoward response to X-irradiation, and tuberous sclerosis. *Am. J. Dis. Child.*, **114**, 617–25, (1967).
8. J. L. Morgan, T. M. Holcomb and R. W. Morrissey, Radiation reaction in ataxia-telangiectasia. *Am. J. Dis. Child.*, **116**, 557–8. (1968).
9. P. N. Cunliffe, J. R. Mann, A. H. Cameron, K. D. Roberts and H. W. C. Ward, Radiosensitivity in ataxia-telangiectasia. *Brit. J. Radiol.*, **48**, 374–376 (1975).
10. J. E. Cleaver, Defective repair replication of DNA in xeroderma pigmentosum. *Nature*, **218**, 652–6 (1968).
11. A. M. R. Taylor, D. G. Harnden, C. F. Arlett, S. A. Harcourt, A. R. Lehmann, S. Stevens and B. A. Bridges, Ataxia-telangiectasia: a human mutation with abnormal radiation sensitivity. *Nature*, **258**, 427–9 (1975).
12. A. M. R. Taylor, Cytogenetics of ataxia-telangiectasia. In B. A. Bridges and D. G. Harnden, eds, *Ataxia-telangiectasia; A Cellular and Molecular Link between Cancer, Neuropathology, and Immune Deficiency.* Chichester: Wiley, 1982, pp. 53–81.
13. D. C. Thrush, Neurological complications of some uncommon cutaneous disorders. In P. J. Vinken and G. W. Bruyn, eds. *Handbook of Clinical Neurology*, vol 38: Neurological manifestations of systemic diseases, part 1. Amsterdam: North-Holland, 1979, pp 1–13.
14. A. D. Andrews, S. F. Barrett and J. H. Robbins, Relation of DNA repair processes to pathological ageing of the nervous system in xeroderma pigmentosum. *Lancet*, **i**, 1318–20 (1976).
15. A. D. Andrews S, F, Barrett and J. H. Robbins, Xeroderma pigmentosum neurological abnormalities correlate with colony-forming ability after ultraviolet radiation, *Proc. natl. Acad. Sci. USA*, **75**, 1984–8 (1978).
16. J. H. Robbins, F. Otsuka, R. E. Tarone, R. J. Polinsky, R. A. Brumback and L. E. Nee, Parkinson's disease and Alzheimer's disease: hypersensitivity to X rays in cultured cell lines. *J. Neurol. Neurosurg. Psychiat.*, **48**, 916–923 (1985).
17. R. E. Tarone, F. Otsuka and J. H. Robbins, A sensitive assay for detecting

hypersensitivity to ionizing radiation in lymphoblastoid lines from patients with Duchenne muscular dystrophy and primary neuronal degenerations. *J. neurol. Sci.*, **65**, 367–81 (1984).
18. J. H. Robbins, A. N. Moshell, S. G. Scarpinato and R. A. Tarone, Cells from patients with olivopontocerebellar atrophy and familial dysautonomia are hypersensitive to ionizing radiation. *Clin. Res.*, **28**, 290A (1980).
19. D. A. Scudiero, A. N. Moshell, R. G. Scarpinato *et al.*, Lymphoblastoid lines and skin fibroblasts from patients with tuberous sclerosis are abnormally sensitive to ionizing radiation and to a radiomimetic chemical. *J. Invest. Dermatol.*, **76**, 234–8 (1982).
20. J. H. Robbins, D. A. Scudiero, F. Otsuka *et al.*, Hypersensitivity to DNA-damaging agents in cultured cells from patients with Usher's syndrome and Duchenne muscular dystrophy. *J. Neurol. Neurosurg. Psychiat.*, **47**, 391–8 (1984).
21. R. R. Weichselbaum, J. Nove and J. B. Little, Skin fibroblasts from a D-deletion type retinoblastoma are abnormally X-ray sensitive. *Nature*, **266**, 726–9 (1977).
22. F. Otsuka, R. E. Tarone, L. R. Seguin and J. H. Robbins, Hypersensitivity to ionising radiation in cultured cells from Down syndrome patients *J. neurol. Sci.*, **69**, 103–112 (1985).
23. C. F. Arlett and W. J. Muriel, Radiosensitivity in Huntington's chorea cell strains: a possible pre-clinical diagnosis. *Heredity*, **42**, 276 (1979).
24 A. E. Harding and K. J. Zilkha, 'Pseudodominant' inheritance in Friedreich's ataxia. *J. med. Genet.*, **18**, 285–7 (1981).
25. P. D. Lewis, J. B. Corr, C. F. Arlett and S. A. Harcourt, Increased sensitivity to gamma irradiation of skin fibroblasts in Friedreich's ataxia. *Lancet*, **ii**, 474–5 (1979).
26. S. Chamberlain and P. D. Lewis, Studies of cellular hypersensitivity to ionising radiation in Friedreich's ataxia. *J. Neurol. Neurosurg. Psychiat.*, **45**, 1136–8 (1982).
27. S. Brennan and P. D. Lewis, Studies of cellular radiosensitivity in hereditary disorders of nervous system and muscle. *J. Neurol. Neurosurg. Psychiat.*, **46**, 1143–5 (1983).
28. S. Chamberlain, W. A. Cramp and P. D. Lewis, Defects in newly synthesised DNA in skin fibroblasts from patients with Friedreich's ataxia. *Lancet*, **i**, 1165 (1981).
29. H. J. Evans, Vijayalaxmi, B. Pentland and M. S. Newton, Mutagen hypersensitivity in Friedreich's ataxia. *Ann. hum. Genet.*, **47**, 193–204 (1983).
30. A. E. Harding, Friedreich's ataxia: a clinical and genetic study of 90 families with an analysis of early diagnostic criteria and intrafamilial clustering of clinical features. *Brain*, **104**, 589–620 (1981).
31. C. F. Arlett and S. A. Harcourt, Survey of radiosensitivity in a variety of human cell strains. *Cancer Res.*, 40, 926–32 (1980).
32. R. R. Weichselbaum, J. Nove and J. B. Little, X-ray sensitivity of fifty-three human diploid fibroblast cell strains from patients with characterised genetic disorders. *Cancer Res.*, 40, 920–25. (1980).
33. R. Cox, G. P. Hosking and J. Wilson, Ataxia telangiectasia. Evaluation of radiosensitivity in cultured skin fibroblasts as a diagnostic test. *Arch. Dis. Child.*, **53**, 386–90 (1978).
34. N. T. Bech-Hansen, W. A. Blattner, B. M. Sell *et al.*, Transmission of in-vitro radioresistance in a cancer-prone family. *Lancet*, **i**, 1335–7 (1981).
35. S. A. Sabovljev, W. A. Cramp, P. D. Lewis *et al.*, Use of rapid tests of cellular radiosensitivity in radiotherapeutic practice. *Lancet*, **ii**, 787 (1985).
36. R. R. Weichselbaum, J. Epstein and J. B. Little, *In vitro* radiosensitivity of human

diploid fibroblasts from patients with unusual clinical responses to radiation. *Radiology*, **121**, 479–82 (1976).

37. S. E. James, C. F. Arlett, M. H. L. Green and B. A. Bridges, Radiosensitivity of human T-lymphocytes proliferating in long term culture. *Int. J. Radiat. Biol.* **44**, 417–22 (1983).
38. G. Harris, W. A. Cramp, J. C. Edwards *et al.*, Radiosensitivity of peripheral blood lymphocytes in autoimmune disease. *Int. J. Radiat. Biol.*, **47**, 689–99 (1985).
39. G. Harris, A. Holmes, S. A. Sabovljev *et al.*, Sensitivity to X-irradiation of peripheral blood lymphocytes from ageing donors. *Int. J. Radiat. Biol.*, **50**, 685–94 (1986).

Radiation and Health
Edited by R. Russell Jones and R. Southwood

14

Cytogenetic Damage: Threshold Effects and Sensitivities

H. J. Evans
Medical Research Council, Clinical and Population Cytogenetics Unit, Western General Hospital, Edinburgh, UK

INTRODUCTION

In any discussion of the longer term effects of exposure of man to ionizing radiations, the two principal concerns are the possible induction of genetic damage that may be transmitted to descendant offspring and future generations, and the induction of cancers in those who are actually exposed. These two biological effects are in reality closely connected for there is, of course, abundant evidence that one or more mutational changes may be a prerequisite for the development of many human cancers, and that an alteration in genetic expression may be a requirement for all.

The kinds of genetic damage induced by radiations encompass a whole variety of changes ranging from very small alterations in DNA base composition to larger changes consequent to the breakage of chromosomes and leading to loss or exchange of chromosome segments, or to even larger changes reflected in the gain or loss of whole chromosomes. The smaller changes are loosely referred to as point mutations and the larger as chromosomal mutations. If we X-irradiate human somatic cells in culture and assay for the kinds of mutation induced it turns out that the majority are of the chromosomal type, moreover a significant proportion of these mutations can be readily observed as alterations in chromosome structure that are quite easily visible under the microscope. It is these visible cytogenetic changes that I shall be concerned with in my discussion.

Chromosomal mutations arise spontaneously in both human germ cells and somatic cells and they contribute in large measure to inherited human disease and, to an as yet unmeasured extent, acquired disease. Chromosomal mutations in germ cells are responsible for a range of congenital abnormalities in the newborn, where they affect 1 in every 170 live births, and are responsible for up to 50 per cent or so of early abortions;[1] they also appear to be important mutational changes in somatic cells where specific mutations are associated with certain forms of cancer such as chronic myeloid leukaemia, where there is a specific change involving chromosomes 9 and 22, or acute lymphocytic leukaemia, where chromosomes 14 and 18 may be involved in a translocation.[2] The kinds of mutation that we see arising spontaneously are also induced on exposure of our cells to mutagens, but of course their frequencies are much increased. How can we measure these frequencies?

HUMAN PERIPHERAL BLOOD LYMPHOCYTES AND THE MEASUREMENT OF CYTOGENETIC DAMAGE

The information that we have on the cytogenetic response of human germ cells to radiations is minimal; on the other hand, a great deal of information has been acquired on the cytogenetic response of human somatic cells to a whole range of mutagens including ionizing radiations. This has been made possible by the development of simple techniques for the short-term, two-day, culture of peripheral blood lymphocytes under conditions which trigger these normally non-dividing cells to enter mitosis and reveal their chromosomes. One millilitre of human peripheral blood contains over 1 million small lymphocytes, so that from one drop of blood we can obtain large numbers of mitotic cells for chromosome analysis. Moreover, we can take blood samples from people before and after they have been exposed to a mutagen, such as X-rays, and measure the number of damaged chromosomes present in their blood cells before and after their exposure. In addition to such *in vivo* studies, we can also undertake *in vitro* studies by taking peripheral blood from a normal individual and exposing the blood cells to mutagens outside the body prior to culture. Before examining the sensitivity of our chromosomes, and possible threshold effects, to exposure to radiations and other mutagens, let us first ask what is the background frequency of chromosomal mutations that we find in healthy people not knowingly and unduly exposed to environmental mutagens?

If we examine blood chromosomes from a population of healthy, young, non-cigarette smoking adults then we find that the frequency of cells that are missing or have gained a chromosome, i.e. have 45 or 47 rather than the normal 46 chromosomes, is of the order of 1 per cent. If we examine the cells in detail then we find that around 1 per cent have chromosome breakage or other structural rearrangements, and for a specific type of rearrangement such as a dicentric chromosome the incidence is around 1 in 1–2000 cells. If we examine cells from older, healthy individuals not unduly exposed to known mutagens, then these frequencies are increased and particularly so for the cells with chromosome gains or losses which may be five or six times higher in 60-year-olds relative to 20-year-olds.[3]

From what I have said it is evident that the background frequencies of chromosomal mutations in our blood cells are relatively high and easy to measure. Although I should mention that a detailed analysis of a metaphase cell, involving identifying and counting aberrations, takes time and skill and it is almost a week's work for a trained technician to score around 1000 cells. This ease of obtaining blood cells and of scoring chromosome aberrations, coupled with the fact that exposure to most human mutagens and all ionizing radiations results in dose-related increases in aberration frequencies, has led to the use of chromosomal mutation assays in blood cells as the only practical measure to date of monitoring exposure of people to mutagens, and particularly to ionizing radiations, and to provide us with some kind of biological dosimeter.[4] Let us then turn to consider the response of our chromosomes to ionizing radiations.

SENSITIVITY AND THRESHOLDS IN *in vitro* STUDIES WITH IONIZING RADIATIONS

If we expose human peripheral blood lymphocytes to X-rays, or other ionizing radiations, *in vitro* then we observe an increased aberration frequency which is dose dependent. The shape of the dose–response curve depends on radiation quality. With X-rays the dose–response curve is clearly curvilinear with the aberration frequency increasing at approximately the square of the dose at

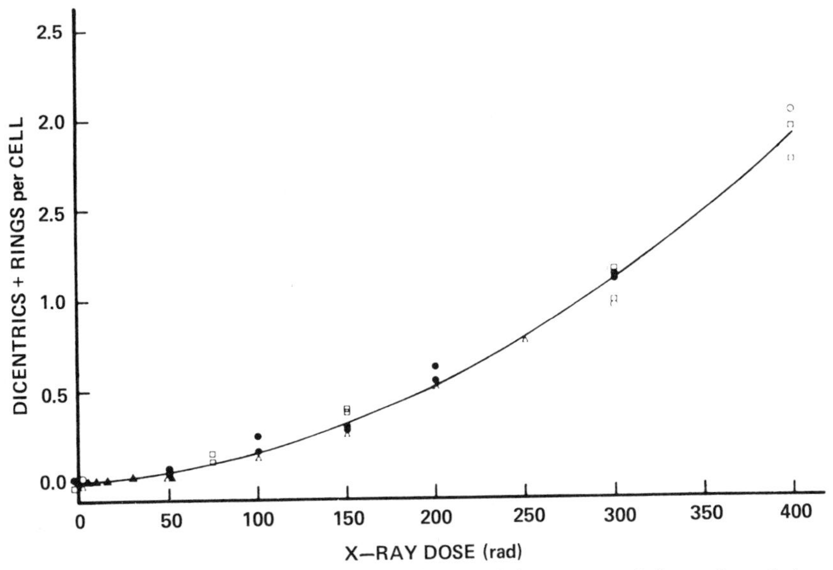

Figure 1 Relationship between X-ray dose and frequency of dicentric and ring chromosome aberrations in human peripheral blood lymphocytes exposed to X-rays *in vitro*. Combined data from five experiments.

high doses, but being approximately linear at low doses (Figure 1). Moreover, the shape of the curve is dose-rate dependent, the longer the exposure time, i.e. the lower the dose rate, the lower the aberration yields. In other words there is some process of inherent repair of radiation damage that reduces the genetic effects.

At a dose of around 25 rads or so given at a high dose rate we see approximately 1 dicentric aberration per 100 cells whereas our normal background frequency is around 1 in every 1–2000 cells. So our chromosomes are sensitive and this led us to ask, some years ago, whether we could detect the effects of exposures at lower doses. Our experiment[5] is summarized in Figure 2. We scored 1500 cells per dose point, but despite this large effort were unable to demonstrate any effects below doses of around 10 rads or so. More recently we have collaborated with thirteen other laboratories in an international endeavour to examine the response at low doses and this exercise involved the counting of tens of thousands of cells. The results (Figure 3) demonstrate that at doses below 5 rads there is no detectable increase in aberration frequency.[6]

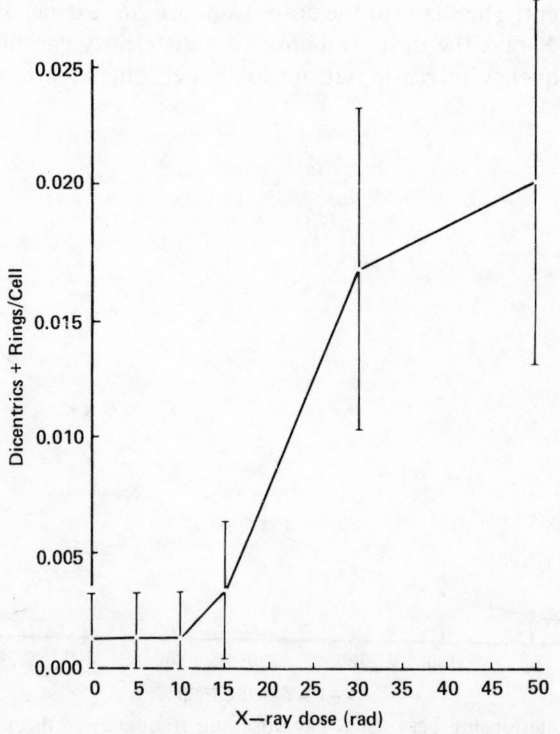

Figure 2 Yield of dicentric and ring aberrations per cell in human lymphocytes exposed to low doses of X-rays *in vitro*.

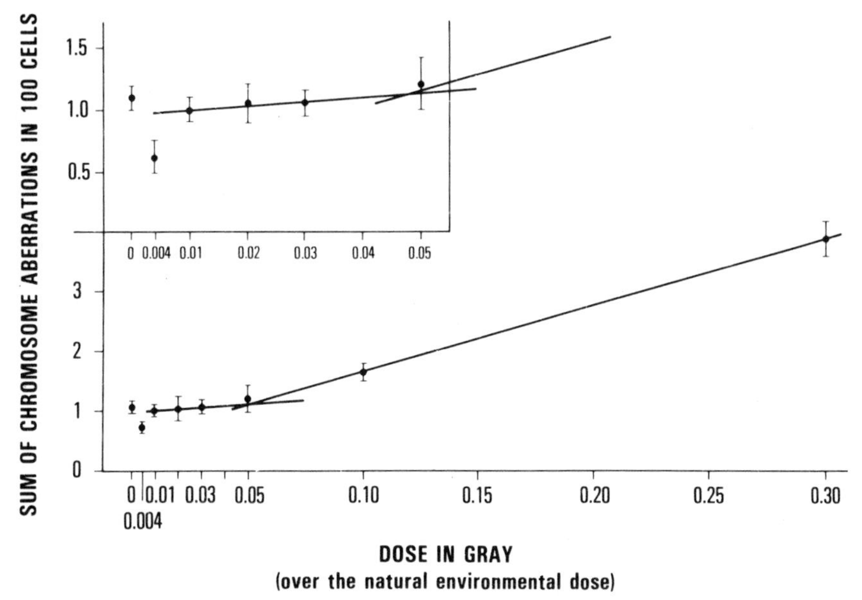

Figure 3 Total aberration frequencies in human lymphocytes exposed to low doses of X-rays *in vitro* (from ref. 6).

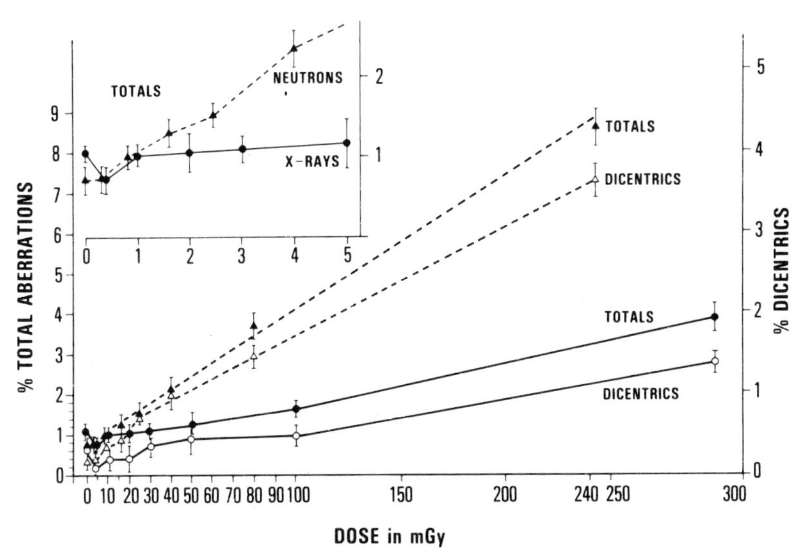

Figure 4 Total aberration frequencies in human lymphocytes exposed to low doses of 240 kV X-rays or 14 MeV fast neutrons *in vitro* (from ref. 7).

The results with high LET radiations, such as fast neutrons or alpha particles, are different from those obtained with low LET radiations, such as X-rays or γ-rays, in three respects. First, for any given level of damage within a biologically meaningful range, high LET radiations are more efficient in damaging than low LET radiations. Second, the relationship of dose versus effect is linear and not curvilinear. Third, there is no dose rate effect, a dose given over a period of many hours is just as efficient as a dose given over a few minutes. This would imply the absence of repair or a threshold. Again the recent international collaborative study has addressed this question and the results (Figure 4) indicate no deviation from linearity in dose response down to levels of around 1 rad.[7]

Let me now turn to ask what information do we have on *in vivo* exposures?

SENSITIVITY AND THRESHOLDS IN *IN VIVO* STUDIES WITH IONIZING RADIATIONS

Some years ago, and as part of a study on the possible therapeutic use of low dose whole-body radiation exposure in cancer therapy, we analysed chromosomes in blood cells of cancer patients immediately prior to, and following, exposure to 25 rads of 2 MeV X-rays. The results showed that we

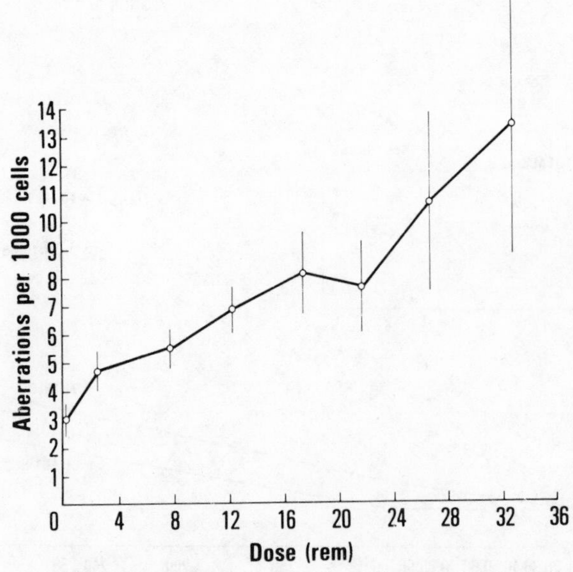

Figure 5 Total aberration frequencies in peripheral blood lymphocytes of nuclear dockyard workers exposed to cumulative doses (largely γ rays) of less than 5 rem per annum over a 10 years period.

could detect around a 15-fold increase in aberration frequency as a consequence of this exposure in these patients. This finding, and certain others, prompted us to undertake a large ten years' study on a group of occupationally exposed dockyard workers who were exposed to radiations, largely gamma-rays, within the occupational limits of 5 rads per annum. We took blood samples from over 200 men before they were ever occupationally exposed to radiation, and then sampled blood cells from them at six-monthly intervals over a period of ten years. All the slides were coded and scored blind and the total analysis was not performed until the study had run for ten years.

We scored a very large number of cells and the results[8] show that with increasing accumulated dose there is a detectable increase in aberration yield (Figure 5)—although no workers were exposed above the occupational limits. If we consider just the dicentrics and ring aberrations we see: (i) that there is no threshold, (ii) that there is an approximately linear dose response and (iii) that we observe significant increases in aberration frequency at doses clearly well below the accepted occupational limits.

CYTOGENETIC EFFECTS OF RADIATIONS AS COMPARED WITH OTHER ENVIRONMENTAL MUTAGENS

Radiations are, of course, but one class of mutagenic hazards and there are vast numbers of mutagenic and carcinogenic chemical agents in the environment which we would expect would induce chromosomal and gene mutations in man. One such agent is cigarette smoke. Cigarette smoking may be responsible for almost one-third of induced cancers in developed countries and a few years ago we and others showed that cigarette smoke condensate is a potent mutagen to human cells exposed in culture and its potency is such that we would expect to see effects on human chromosomes exposed *in vivo*.

In a double blind study we analysed chromosomes in blood lymphocytes from matched smokers and non-smokers, with the smokers smoking on average ten cigarettes a day. Our findings (Table 1) show a highly significant

Table 1 Chromosomal aberrations in blood lymphocytes of cigarette smokers and non-smokers

	Non-smokers	Smokers
Total subjects	41	55
Mean age (yr)	45.8	48.2
Total cells	4100	5500
Gaps	92 (2.24)	171 (3.11)
Deletions	36 (0.87)	56 (1.02)
Exchanges	14 (0.34)	46 (0.84)
Cells with damaged chromosomes	91 (2.21)	176 (3.20)

increase in aberration frequencies in these blood cells taken from light smokers and larger increases have been observed in heavier smokers.[9] Light smokers would appear to show a level of genetic damage that is roughly equivalent to what we would expect of a radiation worker receiving cumulative exposures of around 1–2 rads per year.

In addition to populations of cigarette smokers, various other populations occupationally exposed to known or suspected mutagens have been studied and have shown elevated aberration frequencies, but I will not discuss all of these here. I would, however, like to draw your attention to a recent study by Leonard and his colleagues,[10] who analysed peripheral blood lymphocytes from three groups of Belgian workers: office workers; workers in conventional fossil (coal, oil, gas) fuelled power plants and workers in two nuclear power plants subject to exposure well within the maximum permitted 5 rads per annum. Smoking habits, duration of work at the plants, and radiological examinations were all analysed separately and although there were various minor effects of these three variables, they did not detract from the overall comparisons between the three groups. The results on aberration frequencies, summarized in Table 2, clearly show that there is a small but significant, and consistent, increased frequency of chromosome damage in the blood lymphocytes of the power workers as compared with office controls, and that the fossil-fuelled power workers had more chromosome damage than the nuclear power workers. The increases are small, but significant, and indeed are not surprising for it is well established that the burning of fossil fuels results in the production of highly mutagenic particles and that the fly-ash from fossil-fuelled power stations is a very potent mutagen.[11]

It has been suggested that some of the mutagenicity associated with the burning of fossil fuel is a consequence of the release of radon, and indeed it

Table 2 Chromosomal aberrations (%) in blood lymphocytes of employees from fossil-fuelled and nuclear-power plants (data from ref. 10)

	Controls	Nuclear power workers	Fossil fuel power workers
No. of subjects	23	89	49
No. of cells scored	11 500	44 500	24 500
Cells with aberrations	1.23	1.62*	1.98*
Chromatid aberrations:			
gaps and breaks	0.80	1.04*	1.27*
exchanges	0.05	0.04	0.07
Chromosome aberrations:			
fragments	0.24	0.38*	0.44*
dicentrics	0.11	0.16	0.22

Table 3 Chromosomal aberrations in human lymphocytes cultured in the presence of smog extracts filtered from given volumes of city air (data from ref. 13)

AIR (m^3)	Cells scored	Chromosome breaks	Chromatid breaks	Total breaks
Control (DMSO)	300	4	1	5
1.87 m^3	300	9	7*	16*
3.74 m^3	300	18*	2	20*

has been shown that excluding the radiation we receive as a consequence of medical practice, the largest exposure that we receive over and above normal background levels within the UK is a consequence of the radon released by the burning of fossil fuel.[12] Nevertheless, the amount of radiation released in this way is still very small and obviously far too small to be responsible for the chromosome damage that we see in the fossil fuel power workers.

Exhaust fumes from motor vehicles and city smog are also mutagenic and in some cases their effects are readily detectable in human cells exposed in culture to extracts from filtered emissions involving relatively small volumes of air. For example, Hadnagy and colleagues[13] recently showed that a significant increase in sister chromatid exchange was readily demonstrated in dichloromethane extracts of deposits on glass-fibre filters from one-half of a cubic metre of air from city smog in the heavily industrialized Rhine–Ruhr area of Düsseldorf. These authors also observed increased frequencies of chromosome aberrations in human peripheral blood cells exposed to filtrates from two to three cubic metres of city smog (Table 3) and their findings confirm and extend a large number of earlier reports demonstrating the mutagenic activity of fly-ash in a variety of systems *in vitro*.[14]

CONCLUSIONS

I should like to conclude by summarizing some of the principal findings that are relevant to our concerns regarding the response of our chromosomes to ionizing radiations.

(1) It is evident that the spontaneous rate of chromosomal mutations in our cells is high. It is likely that much of this spontaneous mutation is a consequence of mutagenic agents that are produced intracellularly as normal metabolic by-products, for example various reactive oxygen species are highly mutagenic and they are produced (and usually inactivated) as a consequence of normal metabolic processes in normal cells. It is also possible that a proportion of the spontaneous mutations that we observe is a consequence of unavoidable exposure to mutagenic agents in the environment. The relative

contributions of these intrinsic and extrinsic factors to our mutation load is unknown.

(2) Our chromosomes are sensitive to radiation exposure and we can detect the effects of quite low doses of around the equivalent of 5 rads or so of X-rays. In this context I should emphasize here that the variation in response at low dose levels between different individuals is relatively small: most people, and the cells from most people, show much the same response. As with all biological phenomena, however, there is a range and some people may be more sensitive than others. At the extreme an enhanced response to ionizing radiations is found in certain individuals with rather uncommon inherited diseases which are associated with a reduced ability to repair mutagen-damaged DNA, but these are relatively rare and the overall range of sensitivities to low doses within the population is not large. It does not follow, however, that a similar relatively limited range in response at low exposure levels is to be found for chemical mutagens, where inherent individual differences in metabolism may have very marked effects on the response to a variety of chemical agents.

(3) The relationship between dose and chromosome damage with densely ionizing (high LET) radiations is linear and dose rate independent. We can detect the effects of such radiations at low doses down to around 1 rad and there is no evidence for a threshold or for cellular repair. This contrasts with sparsely ionizing X or γ-rays (low LET) where the dose response is curvilinear and dose rate dependent. Damage from these radiations is subject to repair and little or no effects can be demonstrated at doses below 5 rads.

(4) Finally, in any discussion concerned with environmental hazards, we must not forget that ionizing radiations comprise one category of physical mutagens that may be damaging and must be subject to control. Unfortunately, we often tend to minimize, or indeed even ignore by passive acceptance, other chemical agents in our environment which may be present in greater abundance and which in many instances may be far more powerful than ionizing radiations and present us with far greater hazard.

REFERENCES

1. H. J. Evans, Genetic damage and cancer, in *Genes and Cancer* (ed. J. M. Bishop, J. D. Rowley and M. Greaves), Alan Liss, New York, 1984, pp. 3–18.
2. H. J. Evans, The role of human cytogenetics in studies of mutagenesis and carcinogenesis, in *Genetic Toxicology of Environmental Chemicals, Part A: Basic Principles and Mechanisms of Action* (ed. C. Ramel, B. Lambert and J. Magnusson), Alan Liss, New York, 1986, pp. 41–69.
3. H. J. Evans, Cytogenetic and allied studies in populations exposed to radiations and chemical agents, in *Assessment of Risk from Low-Level Exposure to Radiation and Chemicals. A Critical Overview* (ed. A. D. Woodhead, C. J. Shellabarger, V. Pond and A. Hollaender), Plenum, New York, 1985, pp. 429–51.

4 D. C. Lloyd, An overview of radiation dosimetry by conventional cytogenetic methods, in *Biological Dosimetry* (ed. W. G. Eisert and M. L. Mendelsohn), Springer-Verlag, Berlin, 1984, pp. 3–14.
5. M. Kucerova, A. J. B. Anderson, K. E. Buckton and H. J. Evans, X-ray-induced chromosome aberrations in human peripheral blood leucocytes: the response to low levels of exposure *in vitro. Int. J. Radiation. Biol.*, **21**, 389–96 (1972).
6. J. Pohl-Ruling, P. Fischer, O Haas, G Obe, A. T. Natarajan, P. P. W. Van Buul, K. E. Buckton, N. O. Bianchi, M. Larramendy, M. Kucerova, Z Polikova, A. Leonard, L. Fabry F. Palitti, T. Sharma, W. Binder, R. N. Mukherjee, and U. Mukherjee. Effect of low-dose acute X-irradiation on the frequencies of chromosomal aberrations in human peripheral lymphocytes *in vitro. Mutation Res.*, 71–82 (1983).
7. J. Pohl-Ruling, P. Fischer, K. E. Buckton, R. N. Mukherjee, W. Binder R. Nowotny, W. Schmidt, N. O. Bianchi, P. P. W. Van Buul, A. T. Natarajan, L. Fabry, A. Leonard, M. Kucerova, D. Lloyd, U. Mukherjee, G. Obe, F. Palitti, and T. Sharma. Comparison of dose dependence of chromosome aberrations in peripheral lymphocytes at low levels of acute in vitro radiation with 250 kV X-rays and 14 MeV neutrons, in *Biological Effects of Low-Level Radiation*, IAEA, Vienna, 1983, pp. 171–84.
8. H. J. Evans, K. E. Buckton, G. E. Hamilton and A. Carothers, Radiation-induced chromosome aberrations in nuclear-dockyard workers. *Nature*, **277**, 531–4 (1979).
9. H. J. Evans, Cigarette smoke induced DNA damage in man, in *Progress in Mutation Research, Vol. 2* (ed. A. Kappas), Elsevier/North-Holland, Amsterdam, 1981, pp. 111–28.
10. A. Leonard, G. Deknudt, E. D. Leonard and G. Decat, Chromosome aberrations in employees from fossil-fuelled and nuclear power plants. *Mutation Res.*, **138**, 205–12 (1984).
11. A. Leonard and E. P. Leonard, Evaluation of the mutagenic potential of different forms of energy production. *Sci. Total Environ.*, 29, 195–211 (1983).
12. United Nations Scientific Committee on Effects of Atomic Radiation. Report to the General Assembly. Annex C, pp. 107–11, New York, 1982.
13. W. Hadnagy, N. H. Seemayer and R. Tomingas, Cytogenetic effects of airborne particulate matter in human lymphocytes *in vitro. Mutation Res.*, **175**, 97–101 (1986).
14. M. Moller, and I. Lafheim, Mutagenicity of air samples from various combustion sources. *Mutation Res.*, **116**, 35–46 (1983).

Radiation and Health
Edited by R. Russell Jones and R. Southwood

15

Cancer Incidence and Background Levels of Radiation

Y. UJENO
Department of Experimental Radiology, Faculty of Medicine, Kyoto University, Kyoto, Japan

Analysis of dose–effect relationships was an excellent method for establishing the target theory[1] in radio-biology, though the dose or dose rate ranges used in early studies were relatively high. Both ranges have been lowered with the development of radio-biology, as the analyses of dose–effect relationships have been used to study the risk evaluation of stochastic effects, especially, radiation-induced cancer.[2] The development of highly sensitive measurement techniques of radiation effect and techniques to analyse a huge amount of material such as flow cytometry are needed for this kind of experiment. As the dose and dose rate range are lowered, these requirements become more critical.

Although experimental studies produce useful results, we still require *in vivo* data on humans to evaluate the risk to man. We already have much data on radiation effects from the people exposed in nuclear accidents, atomic bomb survivors, and patients receiving radiation therapy.[3] Similarly much epidemiological data is available on populations who are occupationally or medically exposed to ionizing radiation.[4] However, most of the world population are exposed to even lower doses as a result of background radiation, and though individual doses are small, the total dose represents more than half the whole population dose to mankind. Examination of normal populations in areas with high and low levels of background radiation offers a powerful method of assessing cancer risk in man.[5,6]

The present chapter compares the relationship between the exposure rate of external background radiation in various areas of Japan and the cancer incidence in these areas. There are several reasons for carrying out such a study in Japan. The population density in Japan is relatively high, the genetic background of the population is relatively homogeneous, and social and economical status of the Japanese people are equivalent, and the background radiation level for each prefecture has already been reported.[7] However, the standard deviation of the mean exposure rates for each prefecture was relatively large, compared with the level in cities and towns. This study is therefore limited to cities and towns.

MATERIALS AND METHODS

The data on external background radiation were obtained from the report published by Abe,[7] in which the geographical distribution of measuring points, the techniques of measurement and the instruments of measurement etc. were described in detail. The maximum exposure rate in cities or towns in Japan was $18.9\,\mu R.h^{-1}$ and the minimum was $3.9\,\mu R.h^{-1}$. The maximum rates for cities or towns in Kanagawa and Osaka Prefectures were $7.9\,\mu R.h^{-1}$ and $14.\,\mu R.h^{-1}$, respectively, and the minimum were $3.9\,\mu R.h^{-1}$, and $8..7\,\mu R.h^{-1}$ respectively.

The epidemiological data on cancer incidence were quoted from the reports published by Kanagawa prefectural government,[8] Osaka prefectural government[9] and by the Japanese Society of Obstetrics and Gynaecology.[10] The incidence of cancer is shown as an age-adjusted cancer incidence per 10^5 persons and the world population[11] was used as a control population. The differentiation of cancer was down according to the ICD 9th edition, partly 8th edition.

The age-adjusted cancer incidence of all cancers (ICD 8th revision cord 140–209) in Japan in 1973 was 203.0 for males and 146.1 for females, respectively. The percentage incidence of stomach cancer (ICD 8th revision cord 151), lung cancer (ICD 8th revision cord 162) and liver cancer (ICD 8th revision cord 155, 1978) were 44.7, 11.0 and 6.8 per cent in males, respectively. In females, the incidence of stomach cancer, uterine cancer (ICD 8th revision cord 180–182, 2340) and breast cancer (ICD 8th revision cord 174) was 30.4, 18.9, and 9.5 per cent respectively. The distribution of different kinds of cancers was similar in Kanagawa and Osaka Prefectures. The incidence of chorionic malignant diseases was estimated as approximately 3.1 per 10^3 births (5.4 cases per 10^5 persons) in 1976 in Japan.

In the present report, all analyses on the relationship between the exposure rate of external background radiation and age-adjusted cancer incidence were carried out by Kendall's rank correlation method[12] but not the correlation coefficient method. We adopted 5 per cent as a significance level in the present analyses. Though the so-called doubling dose of the exposure rate of external background radiation can be calculated by means of the correlation coefficient method, it has little relevance as we have no data on individual doses from other sources.

RESULTS

Age-adjusted cancer incidence and exposure rate of external background radiation in Kanagawa Prefecture

The exposure rate of 20 cities and towns in Kanagawa Prefecture which were

Table 1 The relationship between age-adjusted cancer incidence and exposure rate of background radiation in Kanagawa Prefecture for 1970–74

Site	ICD 8th revision cord	Sex	τ	Z	$1 - P(z)$	Note
Oral cavity	140–149	M	0.141	0.156	0.125	
		F	0.444	2.486	0.00656	S*
Oesophagus	150	M	0.000	0.000	0.500	
		F	0.0655	0.392	0.348	
Stomach	151	M	0.0755	0.465	0.322	
		F	0.0269	0.166	0.476	
Intestine	152, 153	M	0.290	1.790	0.0367	S
		F	0.355	2.188	0.0146	S
Rectum	154	M	0.446	2.751	0.00297	S
		F	0.376	2.320	0.0107	S
Liver	155, 1978	M	0.0829	0.496	0.312	
		F	−0.0914	0.564	0.287	
Pancreas	157	M	0.457	2.818	0.00241	S
		F	−0.00538	0.0331	0.488	
Larynx	161	M	0.407	2.357	0.00936	S
		F	−0.263	1.252	0.105	
Lung	162	M	0.296	1.823	0.0343	S
		F	0.0120	0.0717	0.472	
Bone	170	M	−0.0170	0.0920	0.460	
		F	−0.311	1.614	0.0536	
Skin	172 173	M	0.157	0.747	0.472	
		F	−0.0293	0.152	0.440	
Prostate	185	M	0.257	1.438	0.0763	
Breast	174	M	0.000	0.000	0.500	
		F	0.0430	0.265	0.397	
Uterus	180–182, 2340	F	0.0753	0.464	0.322	
Ovary	183	F	0.336	1.945	0.0258	S
Urinary bladder	188	M	0.193	1.121	0.131	
		F	−0.0470	0.272	0.393	
Leukaemia	204–207	M	0.317	1.899	0.0293	S
		F	−0.0484	0.298	0.385	
Others		M	0.0323	0.200	0.424	
		F	0.0305	0.188	0.428	
All cancers	140–209	M	0.253	1.558	0.0605	
		F	0.145	0.895	0.186	

*Significant: $1 - P(z) < 0.0500$.

observed in the present study was 3.9–7.9 μR.h^{-1}. Total cancers were 3759 per year for males and 3323 for females from 1970–74. Table 1 shows the analysed results for 1970–74.

A significant correlation was observed in the age-adjusted cancer incidence in the intestine, the rectum, the pancreas, the larynx, the lung, and leukaemia in males, and of the intestine, the rectum and the ovary in females.

Table 2 shows the results for 1976–8. A significant correlation was observed for the cancer of the intestine, rectum, and pancreas in males and of breast

Table 2 The relationship between age-adjusted cancer incidence and exposure rate of background radiation in Kanagawa Prefecture for 1976–8

Site	ICD 8th revision cord	Sex	τ	z	$1 - P(z)$	Note
Oral cavity	140–149	M	0.0596	0.322	0.374	
		F	−0.126	0.731	0.232	
Oesophagus	150	M	−0.194	1.193	0.130	
		F	−0.248	1.339	0.0901	
Stomach	151	M	0.161	0.994	0.161	
		F	0.0161	0.0992	0.461	
Intestine	152, 153	M	0.537	3.113	0.00093	S
		F	−0.0779	0.466	0.321	
Rectum	154	M	0.355	2.188	0.0140	S
		F	0.197	1.183	0.119	
Liver	155, 1978	M	−0.0359	0.215	0.414	
		F	0.137	0.821	0.206	
Pancreas	157	M	0.367	2.125	0.0170	S
		F	0.174	1.0392	0.149	
Larynx	161	M	−0.0800	0.464	0.320	
		F	−0.566	1.960	0.0250	S(neg.)
Lung	162	M	0.0485	0.299	0.383	
		F	0.150	0.896	0.185	
Bone	170	M	0.0752	0.421	0.337	
		F	0.00145	0.00545	0.498	
Skin	172, 173	M	0.0879	0.457	0.323	
		F	−0.303	1.511	0.0650	
Prostate	185	M	0.0659	0.391	0.348	
Breast	174	F	0.329	1.971	0.0244	S
Uterus	180–182, 2340	F	0.171	1.0520	0.146	
Ovary	183	F	0.212	1.188	0.119	
Urinary bladder	188	M	−0.132	0.788	0.215	
		F	0.125	0.642	0.261	
Leukaemia	204–207	M	−0.187	1.0810	0.140	
		F	0.287	1.660	0.0484	S
Others		M	0.0645	0.398	0.345	
		F	0.0269	0.166	0.434	
All cancers	140–209	M	0.140	0.862	0.193	
		F	0.0323	0.199	0.421	

cancer and leukaemia in females. A negative correlation was observed for laryngeal cancer in females.

For the period 1975–9, the observed population was 35.4×10^5 females, and total cancers were 4773 per year for males and 4059 per year for females.

Table 3 shows the results for 1975–9. A significant correlation was observed for cancer of the intestine, rectum, pancreas and all cancers in males, and for bladder in females. A significant negative correlation was observed for cancer of the oral cavity and larynx in females.

Table 3 The relationship between age-adjusted cancer incidence and exposure rate of background radiation in Kanagawa Prefecture for 1975–9

Site	ICD 9th revision cord	Sex	τ	z	$1 - P(z)$	Note
Oral cavity	140–149	M	0.0537	0.311	0.374	
		F	−0.329	1.906	0.0257	S(neg)
Oesophagus	150	M	0.324	0.199	0.420	
		F	−0.287	1.493	0.068	
Stomach	151	M	0.0108	0.0663	0.436	
		F	−0.0917	0.162	0.435	
Intestine	152, 153	M	0.493	2.947	0.0164	S
		F	−0.113	0.936	0.174	
Rectum	154	M	0.333	2.0552	0.0228	S
		F	0.126	0.753	0.226	
Liver	155	M	0.258	1.541	0.0611	
		F	0.0215	0.133	0.448	
Pancreas	157	M	0.366	2.254	0.0122	S
		F	0.172	1.0608	0.145	
Larynx	161	M	0.262	1.517	0.0655	
		F	−0.418	2.173	0.0117	S(neg.)
Lung	162	M	0.0538	0.331	0.371	
		F	−0.0541	0.333	0.370	
Bone	170	M	−0.162	0.937	0.174	
		F	0.163	0.221	0.412	
Skin	172, 173	M	−0.0419	0.251	0.401	
		F	−0.0685	0.384	0.351	
Prostate	185	M	−0.172	1.061	0.145	
Breast	174	F	0.204	1.259	0.106	
Uterus	179–182, 2331	F	−0.0323	0.199	0.425	
Ovary	183	F	0.246	1.469	0.0721	
Urinary bladder	188	M	0.151	0.928	0.178	
		F	0.330	1.913	0.0280	S
Leukaemia	204–208	M	0.234	1.397	0.0823	
		F	−0.102	0.609	0.274	
Others		M	0.129	0.796	0.214	
		F	−0.140	0.862	0.195	
All cancers	140–208	M	0.290	1.790	0.0367	
		F	−0.0592	0.365	0.356	

Age-adjusted incidence and exposure rate of external background radiation in Osaka Prefecture

The exposure rate of external background radiation in cities and towns in Osaka Prefecture was 8.7–14.4 μR.h^{-1}. The population size was 4.2×10^5 males and 4.3×10^5 females in 1983. Only cancer of the stomach, colon, rectum, liver, pancreas, lungs, breast and uterus were included in the analysis. Other cancers were not analysed due to the lack of epidemiological data in the referred reports.

The incidence of male colonic cancer was significantly correlated with the exposure rate for 1975–9, the incidence of uterine cancer was significant for 1982, and the incidence of male stomach cancer and female breast cancer were significant but negative for 1983. Table 4 shows the results for 1975–9.

A comparison of the results obtained in Kanagawa and Osaka Prefectures

The incidence of intestinal, rectal and pancreatic cancers in males were significantly correlated with the exposure rate for three periods in Kanagawa Prefecture. Laryngeal cancer in females showed a negative correlation not only for 1976–8 but also for 1975–9 in Kanagawa Prefecture.

The results of six periods in Kanagawa and Osaka Prefectures listed in Table 5, show that no cancer is significantly correlated with exposure rate for all six time periods. Five cancers, that is, cancer of oral cavity in females, stomach cancer in Osaka in 1983, laryngeal cancer in Kanagawa Prefecture in males for 1976–8 and 1975–9, and breast cancer in Osaka Prefecture in 1983, were negatively correlated with the exposure rate.

Table 4 The relationship between age-adjusted cancer incidence and exposure rate of background radiation in Osaka Prefecture for 1975–9

Site	ICD 9th revision cord	Sex	τ	z	$1 - P(z)$	Note
Stomach	151	M and F	−0.0300	0.168	0.433	
Colon	153	M	0.330	1.846	0.0320	S
		F	0.0188	0.105	0.146	
Rectum	154	M	0.120	0.671	0.251	
		F	0.0906	0.507	0.306	
Liver	155	M	0.0892	0.500	0.309	
		F	0.0595	0.333	0.370	
Lung	162	M and F	0.0297	0.167	0.433	
Breast	174	F	0.0446	0.250	0.401	
Uterus	179–182, 2331	F	−0.134	0.750	0.227	

Table 5 Relationship between age-adjusted cancer incidence and exposure rate of background radiation analysed in the present study

Site (ICD 9th)[a]	Sex	Prefecture: Kanagawa			Prefecture: Osaka		
		1970–74	1976–8	1975–9	1975–9	1982	1983
Oral cavity	M				—[b]	—	—
(140–149)	F	S		S(neg.)	—	—	—
Oesophagus	M				—	—	—
(150)	F				—	—	—
Stomach	M						S(neg)
(151)	F						
Intestine	M	S	S	S	—	—	—
(152, 153)	F	S			—	—	—
Colon	M	—	—	—	S		
(153)	F	—	—	—			
Rectum	M	S	S	S			
(154)	F	S					
Liver	M						
(155)	F						
Pancreas	M	S	S	S	—		
(157)	F				—		
Larynx	M	S			—	—	—
(161)	F		S(neg.)	S(neg.)	—	—	—
Lung	M	S					
(162)	F						
Bone	M				—	—	—
(170)	F				—	—	—
Skin	M				—	—	—
(172, 173)	F				—	—	—
Breast	M		—	—	—	—	—
(174)	F		S				S(neg)
Uterus (179–182, 2331)	F					S	
Ovary (183)	F	S			—	—	—
Prostate (185)	M				—	—	—
Urinary bladder	M				—	—	—
(188)	F			S	—	—	—
Leukaemia	M	S			—	—	—
(204–208)	F		S		—	—	—
Others	M				—	—	—
	F				—	—	—
All cancers	M			S			
(140–208)	F						

[a] ICD 8th revision cord for 1970–74 and 1976–8.

[b] The epidemiological data are not published yet.

Table 6 The relationship between incidence of chroionic diseases and exposure rate of background radiation in 11 prefectures for 1974–6

	τ	z	$1 - P(z)$	Note
Hydatidiform mole				
per 10^5 population	0.495	2.12	0.0170	S
per 10^3 birth[a]	0.222	0.951	0.170	
Hydatidiform mole and malignant hydatidiform mole				
per 10^5 population	0.679	2.906	0.0139	S
per 10^3 births	0.422	1.791	0.0367	S
Chorionepithelioma				
per 10^5 population	0.129	0.550	0.291	
per 10^3 births	0.349	1.479	0.0694	
All chorionic diseases				
per 10^5 population	0.385	1.650	0.0494	S
per 10^3 births	0.312	1.323	0.0934	

[a] Birth-rate adjusted.

Incidence of chorionic diseases and the exposure rate of external background radiation in 11 prefectures

The relationship between the incidence of chorionic diseases and the exposure rate of external background radiation analysed in 11 prefectures was 6.65–10.87 μR.h^{-1}. The total population size was 1079×10^5, and total births for 1974–6 was 19×10^5. The incidence was not adjusted for maternal age, but was adjusted for birth-rate.

The results in Table 6 show that the incidence of hydatidiform mole and malignant hydatidiform mole per 1000 births correlated with the exposure rate of external background radiation. The incidence of chorionic diseases per 100 000 population, was significant for hydatidiform mole, malignant hydatidiform mole, and all chorionic diseases.

DISCUSSION

The present study was designed to test whether stochastic effects of ionizing radiation can be detected at the low doses provided by background radiation. We also need to know whether the dose–effect relationship fits the L-model or LQ-model, or not at very low dose. However, we do not know the exposure rate of external background radiation for each inhabitant. For this reason, we used Kendall's rank correlation coefficient method to analyse the relation between cancer incidence and exposure rate, but did not calculate the risk directly by a correlation coefficient method.

There are many carcinogenic agents including chemical carcinogens in the human environment. Medical exposure to radiation which has a large population dose equivalent is one such agent. In the present study we did not take into account the internal background radiation because ^{40}K is distributed homogeneously in human bodies worldwide. The mutagenicity of ^{40}K to bacteria has been studied[13] but the carcinogenic effect in humans is not clear.

LET is a factor modifying the dose–effect relationship of external background radiation. This modifying factor was not considered in the present study, because the LET value of external background radiation can be assumed to be approximately 1.00.[14] Recently, the role of oncoviruses in radiation-induced cancer has been discussed.[15] However, this factor could not be included in the present study. Populations with a high susceptibility to ionizing radiation have also been investigated[16,17] but we have no data on the geographical location of such a population in Japan. Thus this factor is not considered here.

The results obtained on Kanagawa Prefecture will be discussed first. We compared the results for 1970–74 and 1975–9. Three kinds of cancers, that is, intestinal cancer, rectal cancer and pancreatic cancer in males were significantly related to the exposure rate in both periods. This suggests that these three cancers have some relation to background radiation. Interestingly these cancers occur in the alimentary system and their relative incidence is less than 4 per cent of all cancers for 1975–9. This suggests that other factors which might be important in the aetiology of stomach cancer, are not closely related to the induction of these cancers. Our study shows that pancreatic cancer in males is significantly related to exposure rate. Such a relationship was not observed in our previous study which involved whole prefectures,[18] nor in the Hanford data.[19] However, pancreatic cancer was present in excess in the Hanford workforce according to some authors.[20] No data on the incidence of renal cancer are available for Kanagawa and Osaka Prefectures.

The present results did not show a significant correlation between radiation and the incidence of leukaemia such as has been observed for example in the atomic bomb survivors.[21] It is possible that leukaemia risk per unit dose becomes less at lower dose.

Male cancer was correlated more significantly with background radiation than female cancer. These male/female differences suggest a hormonal influence in cancer induction at low dose. However, no such effect has been observed previously except in the case of sex linked cancers.[22]

Although Osaka and Kanagawa Prefectures have similar socio-economic conditions, no single cancer was significantly correlated with background radiation in both prefectures for the period 1975–9. However, this may reflect the importance of other carcinogenic influences in the two populations.[23,24]

In both prefectures, some significant negative correlations were observed, but they were not common to both prefectures (see Table 5). Such phenomena have been observed previously,[25] and are difficult to explain in radio-biological terms.[26]

We were fortunate in obtaining epidemiological data on chorionic diseases in Japan. However, the standard deviation of the mean incidence calculated for each prefecture is large. Furthermore, the incidence used here was neither adjusted to age of population nor mother's age. The correlation between the exposure rate and incidence is significant in spite of these uncertainties. As shown in Table 6, the incidence of chorionepithelioma did not correlate significantly with the exposure rate, but appears to be based upon the small number of patients in each prefecture. Chorionic disease results from changes in germ cells as opposed to cancer which involves transformation of somatic cells. The way in which ionizing radiation might induce chorionic disease is not well understood.[27]

Today, we have more data on cancer mortality than cancer incidence in Japan. A detailed analysis of the relation between exposure rate and cancer mortality in Japan has already been made[25] and the relation between personal dose equivalent and cancer mortality.[19] The incidence to mortality rate for all cancers was 1.43, the maximum was 3.12 for breast cancer and the minimum was 1.07 for leukaemia in Osaka Prefecture in 1983. If the ratio is accurate for each cancer, we can use the cancer mortality data to analyse the relationship between cancer incidence and background radiation in more detail.

The following will improve our analysis.

1. To obtain a better assessment of the radiation dose of the inhabitants. The daily life pattern of each person must be examined for this purpose.
2. To know the personal dose equivalent due to medical exposure.
3. To collect the data on age-adjusted cancer incidence in relatively small areas, such as cities and towns for a period of more than five years.
4. To know the geographical distribution of various carcinogenic agents in the environment.

This programme is difficult to achieve but is essential not only for the evaluation of stochastic effects but also to know the mechanisms of biological effects of ionizing radiation at very low dose.

In conclusion we could not observe a definite correlation between cancer incidence and background radiation in all six periods of observation. However, several cancers seem to have a significant relationship with background radiation. The dose–effect relationship at very low dose may differ from that at higher dose rates.

REFERENCES

1. D. E. Lea, *Actions of Radiations on Living Cells*, Cambridge University Press, England, 1946.

2. Robin H. Mole, Dose–response relationships, in *Radiation Carcinogenesis: Epidemiology and Biological Significance*, ed. J. D. Boice, Jr. and J. Fraumeni, Jr., Raven Press, New York, 1984, pp. 403–420.
3. Arthur C. Upton, Historical perspectives on radiation carcinogenesis, in *Radiation Carcinogenesis*, ed. A. C. Upton, R. E. Albert, F. J. Burns and R. E. Shore, Elsevier, New York, Amsterdam, London, 1986, pp. 1–10.
4. Seymour Jablon, Epidemiological perspectives in radiation carcinogenesis, in *Radiation Carcinogenesis: Epidemiology and Biological Significance*, ed. J. D. Boice, Jr. and J. F. Fraumeni, Jr., Raven Press, New York, 1984, pp 1–8.
5. Ademar Freire-Maia, Human genetics studies in area of high natural radiation II. First results of an investigation in Brazil. *An. Acad. Brasil. Cienc.*, **43**, 457–9 (1971).
6. Tao Zufen and Wei Luxin, An epidemiological investigation of mutational diseases in the high background radiation area of Yangiang China. *J. Radiat. Res.*, **27**, 141–50 (1986).
7. S. Abe, Effects to obtain Japanese profile of ambient natural background exposure. *Hoken Butsuri* (*J. Jpn. Helth. Phys. Soc.*), **17**, 169–93 (1982).
8. Kanagawa (Department of Health and Welfare, Kanagawa Prefectüre Government) Kanagawa Cancer Registry, *Report 4* (1981), *Report 6* (1982), *Report 7* (1983).
9. Osaka (Department of Health and Welfare, Osaka Medical Association and Cancer for Adult Diseases, Osaka Prefecture Government), Osaka Cancer Registry, *Report 36* (1983), *Report 39* (1985), *Report 40* (1985b), *Report 41* (1986).
10. Obst. Gynec. (The committee for registration of chorionic disease in Japan), The report of the committee for registration of chorionic diseases in Japan. *Act Obst. Gynec. Jpn.*, **31**, 525–30 (1979).
11. UICC, *Cancer Incidence in Five Continents*, ed. J. Waterhouse, C. Muir, P. Correa and J. Powell, Vol. III (1976).
12. M. G. Kendall, *Rank Correlation Method*, London, Charles Griffin, 1948.
13. D. Gevertz, A. M. Friedman, J. J. Katz and H. E. Kubitschek: Biological effects of background radiation: mutagenicity of ^{40}K. *Proc. Natl. Acad. Sci. USA*, **82**, 8602–5 (1985).
14. UNSCEAR, *UN, Sources and Effects of Ionizing Radiation*, 1977 Report to the General Assembly, UN, New York, 1977, Annex A, III-44.
15. J. F. Duplan, B. Guillemain and T. Astier, Role of virus, in *Radiation Carcinogenesis*, ed. Arthur C. Upton, Roy E. Albert, Fredric J. Burns, and Roy E. Shore, Elsevier, New York, Amsterdam, London, 1986, pp. 71–84.
16 Alfred G. Knudson and Suresh H. Moolgavkar, Inherited influences on susceptibility to radiation carcinogenesis, in *Radiation Carcinogenesis*, ed. Arthur C. Upton, Roy E. Albert, Fredric J. Burns and Roy E. Shore, Elsevier, New York, Amsterdam, London, 1986, pp. 401–411.
17. Elaine Ron and Baruch Modan, Thyroid and other neoplasms following childhood scalp irradiation, in *Radiation Carcinogenesis: Epidemiology and Biological Significance*, ed. J. D. Boic, Jr. and J. F. Fraumeni Jr., Raven Press, New York 1984, pp. 139–51.
18. Y. Ujeno, Relation between cancer mortality or cancer incidence and external background radiation in Japan, in *Biological Effects of Low-level Radiation*, IAEA, Vienna, 1983, pp. 253–62.
19. S. C. Darby and J. Reissland, Low levels of ionizing radiation and cancer—are we underestimating the risk? *Roy. Statist. Soc.*, A **144**, 298–331 (1981).
20. T. F. Mancuso, A. Stewart and G. Kneale, Radiation exposures of Hanford

workers dying from cancer and other diseases. *Health Physics*, **33**, 379–85 (1977).

21. M. Ichimaru, T. Okita and T. Ishimaru: Leukemia, multiple myeloma, and malignant lymphoma, in *Gann Monograph on Cancer Research*, ed. I. Shigematsu and A. Kagan, Japan Sci. Press, Tokyo and Plenum Press, New York and London, 1986, Vol. 33, pp. 113–27.
22. Gilbert W. Beebe, Developments in assessing carcinogenic risks from radiation, in *Radiation Carcinogenesis: Epidemiology and Biological Significance*, ed. J. D. Boic Jr. and J. F. Fraumeni Jr., Raven Press, New York, 1984, pp. 457–66.
23. Ulf Flodin, Mats Fredriksson, Olav Alexson, Bodil Persson, Lennart Hardell, Background radiation, electrical work, and some other exposures associated with acute myeloid leukemia in a case-referent study. *Arch. Environ. Hlth*, **41**, 77–84 (1986).
24. R. J. M. Fry and R. L. Ullrich: Combined effects of radiation and other agents, in *Radiation Carcinogenesis*, ed. Arthur C. Upton, Roy E. Albert, Fredric J. Burns and Roy E. Shore, Elsevier, New York, Amsterdam, London, 1986, pp. 437–54.
25. K. Noguchi, M. Shimizu and I. Anzai, Correlation between natural radiation exposure and cancer mortality in Japan (1) *J. Radiat. Res.*, **27**, 191–212 (1986).
26. T. D. Luckey, Physiological benefits from low levels of ionizing radiation. *Health Physics*, **43**, 771–89 (1982).
27. Y. Ujeno, Epidemiological studies on disturbances of human fetal development in areas with various doses of natural background radiation. II. Relationship between incidence of hydatidiform mole, malignant hydatidiform mole, and chorion-epithelioma and gonad dose equivalent rate of natural background radiation. *Arch. Environ. Health*, **40**, 181–4 (1985).

Radiation and Health
Edited by R. Russell Jones and R. Southwood

16

Childhood Cancers in the UK and their Relation to Background Radiation

G. W. Kneale and A. M. Stewart
Regional Cancer Registry, Queen Elizabeth Medical Centre, Edgbaston, Birmingham, UK

ABSTRACT

This chapter shows the results of including two independent data sets in a study of several factors with cancer associations including background radiation. One data set came from the Oxford Survey of Childhood Cancers (OSCC) and the other from the National Radiological Protection Board (NRPB) and the findings are compatible with background radiation being the single most important cause of juvenile neoplasms. It also emerged that these neoplasms have a strongly clustered distribution. No obvious cause of the clusters was found. However, they had associations with prenatal and postnatal illnesses as well as background radiation. Therefore, since there is mounting sensitivity to infections during the latent phase of leukaemia the cancer clusters might be the result of competing causes of death having an epidemic distribution.

The findings as a whole are compatible with all man-made additions to background (including leakages of radioactivity from a reprocessing plant) adding to the risk of an early cancer death. However, proof that certain leukaemia clusters in the vicinity of two reprocessing plants were caused in this way must await the collection of more epidemiological data.

INTRODUCTION

Recent work has shown that the fetal hazard associated with prenatal x-rays has much the same effect on childhood leukaemias and other types of juvenile neoplasms.[1] However, the risk is considerably greater for first trimester x-rays (which are rare) than third trimester x-rays (which are common)[2] and the ratio of young cases (under three years) to older cases is higher for non-x-rayed than

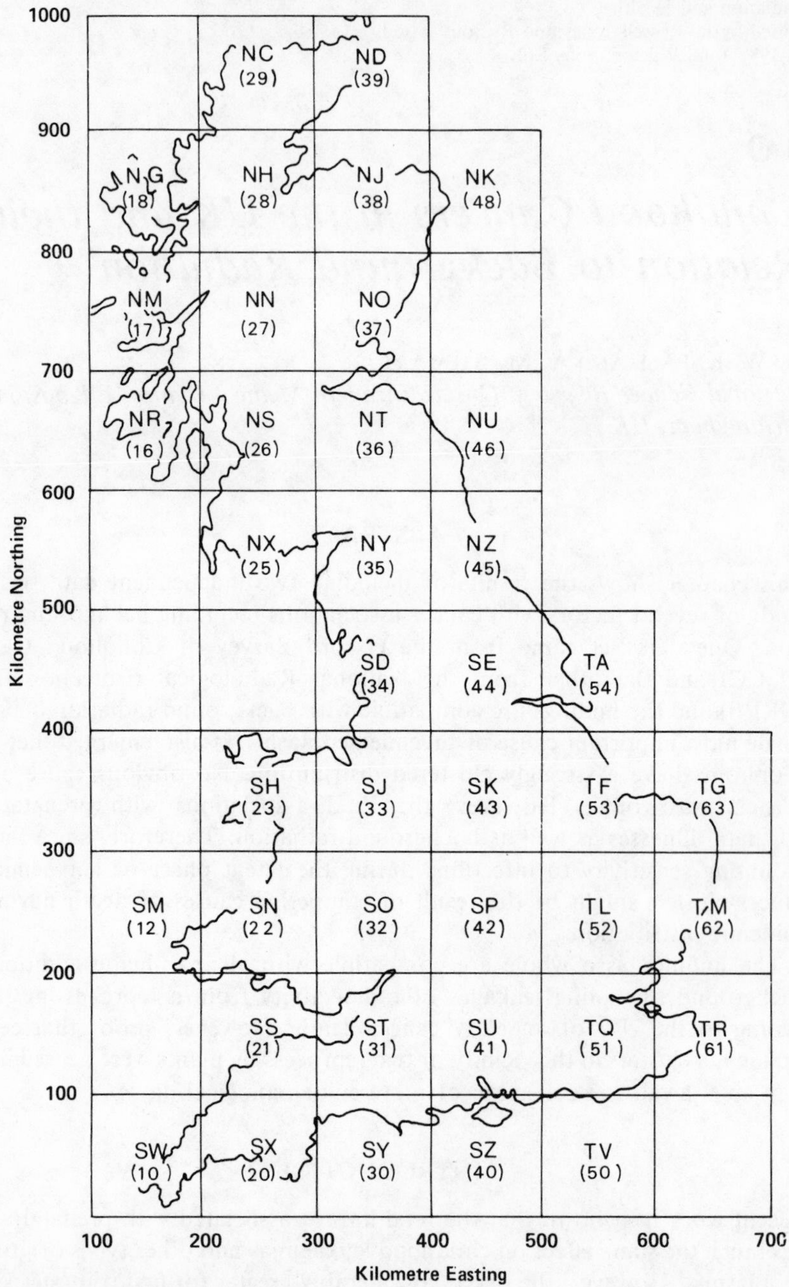

Figure 1 Parts of the national grid included in the Oxford Survey of Childhood Cancers (OSCC data).

x-rayed cases.[3] Therefore, there is a strong presumption that all juvenile neoplasms have fetal origins and include a high proportion of embryomas or cases with near conception origins.[4] The dose to the fetus from a typical obstetric x-ray (about 0.5 rem) is considerably higher than the total *in utero* dose from background radiation (about 0.15 rem). However, the cancer risk

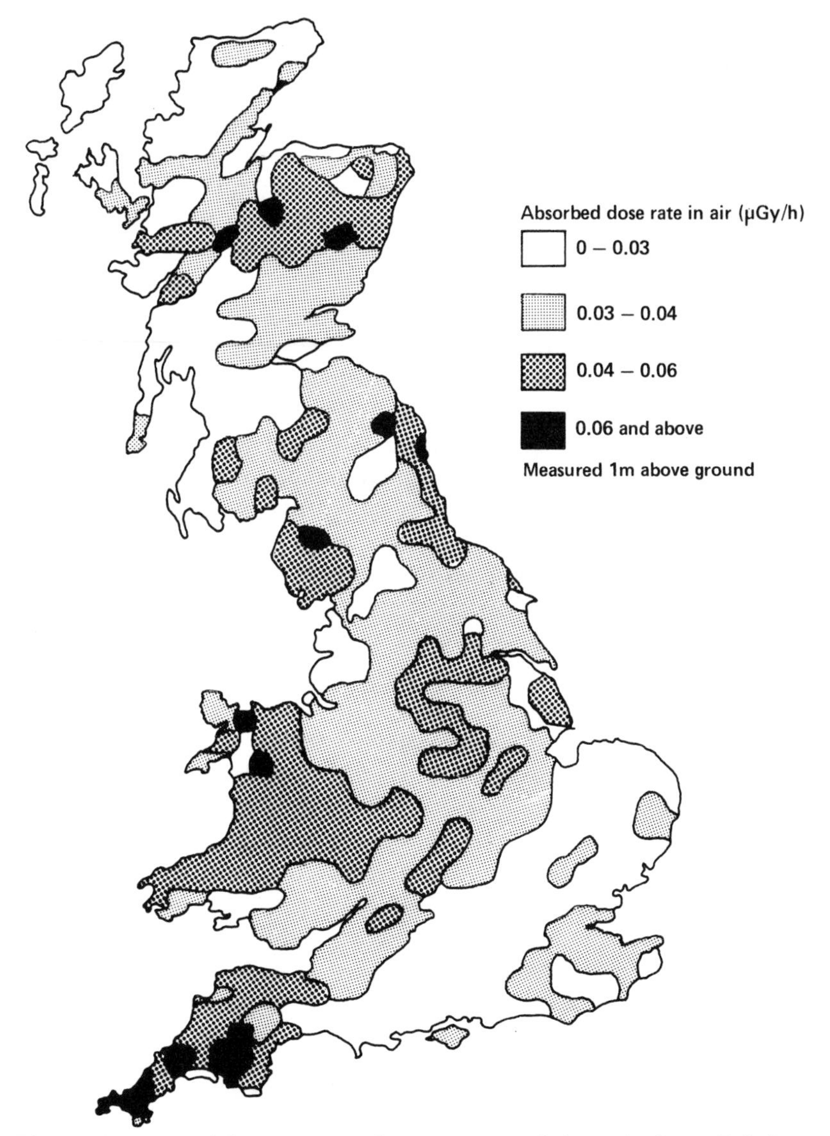

Figure 2 Terrestrial gamma-ray dose rates out of doors in Great Britain (NRPB data).

could be much greater for embryos than mature fetuses, in which case the number of juvenile neoplasms caused by *in utero* exposures to background radiation could be much larger than the number caused by prenatal x-rays.

The first opportunity to test this hypothesis came in 1985. By this time all the regions included an on-going case/control study of juvenile neoplasms (Oxford Survey of Childhood Cancers or OSCC data) (Figure 1) had also been included in a field survey of background radiation by the National Radiological Protection Board. The NRPB data included estimates of the terrestrial gamma-radiation or TGR dose rates for each 10 kilometre square of the national grid (i.e. for 2400 contiguous locations) also the map based on these estimates which is reproduced in Figure 2. Therefore, since OSCC data included the addresses of a large and continuous series of juvenile neoplasms, it was appropriate to include the two data sets in tests of the hypothesis that background radiation is a cause of early cancer deaths.

OSCC DATA

OSCC data included a wide range of information on 22 351 early cancer deaths (0–15 years) and matched controls of these cases. The matching factors were sex, date of birth and region, and for 1797 regions (i.e. boroughs, urban areas and rural districts), there were annual numbers of live births from 1944 to 1979 and annual numbers of cancer deaths from 1953 to 1979. Therefore, in each LA region there were 36 birth cohorts with cancer follow-up periods ranging from 16 years (1953–68 births) to one year (1979 births).

Table 1 Regional birth cohorts with stated numbers of cancer deaths

Cancer deaths per cohort nos	Regional birth cohorts nos	Cancer deaths
0	22 853	0
1	5 779	5 779
2	1 922	3 844
3	813	2 439
4	423	1 692
5	275	1 375
6	180	1 080
7	116	812
8	81	648
9	73	657
10+	81	4 025
1–10+	9 943	22 351
Total	32 796	22 351

By the time each LA region had been given a single grid location (centred on a 10 km square) the 40 large (100 km) squares in Figure 1 were divisible into 911 'districts' with cancer mortality rates (OSCC data) and TGR dose rates (NRPB data). Urban districts were, on average, less extensive than rural districts, but even in the most sparsely populated district there was a continuous series of live births. Therefore there were 36 × 911 or 32 796 'regional birth cohorts' contributing to 347 564 863 person years of follow-up. Only in the most densely populated districts was there a continuous series of cancer deaths and for nearly 3000 of the regional birth cohorts the follow-up period was less than two years. Therefore, the number of cohorts contributing to 22 351 cancer deaths (9943) was much smaller than the number contributing to person years of follow-up (Table 1).

TESTS OF THE BACKGROUND RADIATION HYPOTHESIS

It was first necessary to discover whether the 22 351 cancer deaths had a random or non-random distribution. This point was settled by including all the 32 796 regional birth cohorts which had contributed to the 347 567 826 person years of follow-up in a Poisson heterogeneity test which recognized 911 space units (i.e. districts) and 36 time units (i.e. calendar years of birth). For these units, either separately or combined, there was firm rejection of the random distribution hypothesis. Therefore some of the regional birth cohorts must have experienced an excessive number of early cancer deaths (so-called clusters) and some must have had the opposite experience (i.e. non-clusters or exceptionally lengthy gaps between cases).

The extent of the clustering (within the 9943 cohorts contributing to the cancer deaths) is shown in Table 2. Only these cohorts were eligible for a test

Table 2 Regional birth cohorts with excessive numbers of cancer deaths

Expected number of cancer deaths per cohort[a]	Total	Regional birth cohorts with excessive numbers of cancer deaths[b]	
		No.	%
Under 4	8514	271	3.2
4–5	698	134	19.2
6–7	296	45	15.2
8–9	154	20	13.0
10+	281	47	16.7
Total	9943	517	5.2

[a] Excluding cohorts with no cancer deaths (see Table 1). The greater the expected number the higher the level of population density.

[b] An excessive number of cancer deaths = a cancer cluster or an observed number greater than expected at a 5 per cent Poisson significance level (see Tables 7 to 9).

of whether the observed number of deaths was greater than expected at a 5 per cent Poisson significance level. But even so there were 517 cohorts which exceeded this limit. The cluster proportion was higher for large (urban) cohorts with long follow-up periods (i.e. 1423 cohorts with at least four expected deaths) than for small (rural) cohorts or cohorts with a short follow-up period (i.e. the 8515 remaining cohorts). Therefore, there were clearly two factors influencing cluster detection, namely, population density and follow-up periods.

For cancers with fetal origins, a clustered distribution could be caused in one of two ways: either by *in utero* exposure to an unusually hazardous event (positive effect of exposure to carcinogens) or by *in utero* or postnatal involvement in epidemic diseases (negative effect of competing causes of death).

Earlier analyses of OSCC data had shown, first, that during the interval between cancer induction and diagnosis (pre-onset period) of juvenile neoplasms there is progressive loss of immunological competence[5,6] and, secondly, that as a result of this early change, cases of preleukaemia eventually became so infection sensitive that they are over 300 times as likely to die from respiratory infections as normal children.[7] Therefore, when seeking epidemiological evidence of juvenile cancer effects of background radiation in OSCC data, it is clearly not enough to compare regional cancer mortality rates with radiation dose levels. There must also be control for possible effects of intercurrent infections and infection-related factors such as sex, maternal age, sibship position, social class and population density.

Table 3 Cancer death rates for eight nationwide levels of TGR dose rates

TGR dose rate (nGy/h)	Person years (PY) of follow-up	Cancers[a]	
		Nos	Rate (per 10^5 P/Y)
12–19	5 103 116	362	7.09
20–24	45 687 955	2 973	6.51
25–29	87 605 332	5 664	6.47
30–34	69 027 813	4 334	6.28
35–39	73 368 981	4 826	6.58
40–44	56 315 974	3 548	6.30
45–49	4 399 802	255	5.80
50–82	6 055 863	389	6.42
12–82[b]	347 564 836	22 351	6.43

[a] In this table there are 2 936 633 person years and 50 cancer deaths not included in Table 6 (see text).
[b] Average 32 nGy/h (for all natural background radiation the corresponding dose rate would be either 120 nGy/h or 0.105 rad per annum).

Therefore, the Poisson heterogeneity test of OSCC data was followed, first, by uncontrolled comparisons between cancer rates and TGR dose levels at national level (Tables 3 and 4). Then the whole country was divided into 40 separate zones corresponding to the grid squares in Figure 1 (Tables 5 and 6) and, finally, the regression analysis described in the Appendix was applied to twelve factors with possible cancer associations including prenatal x-rays and TGR dose rates (Tables 7, 8 and 9).

Table 4 Cancer death rates and TGR dose rates for three sections of the national grid

Subdivisions of the national grid	Person years of follow-up	Cancers		TGR dose rate (nGy/h)
		Nos	Rates	
West	69 731 256	4 281	6.14	37.1
Centre	145 543 857	9 371	6.44	34.0
East	132 289 723	8 699	6.58	31.3
All regions	347 564 836	22 351	6.43	33.9

Table 5 Cancer rates and TGR dose rates for one densely populated region measuring only 100 km^2 (see TQ 51 in Figure 1)

TGR dose rate (nGy/h)	Districts[a] (No.)	Person years of follow-up	Cancer deaths	
			No.	Rate
21	3	930 053	50	5.376
22	5	8 154 411	575	7.051
23	1	7 241 573	375	5.178
24	6	9 020 555	589	6.530
25	2	2 052 749	157	7.648
26	5	10 772 972	652	6.052
27	6	6 417 647	449	6.996
28	4	12 229 611	777	6.353
29	8	6 722 945	504	7.497
30	6	2 579 391	236	10.045
31	3	657 439	39	5.932
32	4	746 469	54	7.234
33	4	1 729 740	140	8.094
34	3	989 897	47	4.748
35	1	330 484	40	12.103
27.8	61	70 575 936	4684	6.637
		Regression coefficient	+1.299	(10^{-6})

[a] see text.

Table 6 Cancer rates and TCR dose rates for 36 regions corresponding to 100 km^2 of the national grid

National grid regions[a]		District P/Y of follow-up		Cancer deaths	TGR dose rates		Dose response	
Specifications	Districts	Rank	Average (10^3)	(1935–79)	Lowest	Highest	β	t value
TQ 51	61	1	1157	4684	21	35	+ 1.30	4.00
SJ 33	53	2	686	2207	29	42	+ 0.07	1.94
NZ 45	29	3	646	1109	30	45	+ 1.19	1.91
NS 26	39	4	569	1281	23	35	+ 0.07	0.07
SP 42	53	5	469	1649	23	48	− 0.33	0.97
SK 43	54	6	457	1587	30	45	− 0.63	1.48
SE 44	49	7	418	1318	28	43	− 0.71	1.30
SD 34	42	8	408	1075	25	44	+ 0.95	1.08
SU 41	47	9	370	1241	13	37	− 0.07	0.13
TA 54	15	10	351	318	28	39	+ 2.33	2.07
ST 31	50	11	293	928	21	46	− 0.46	0.92
SO 32	48	12	267	838	35	49	+ 0.96	1.15
TL 52	47	13	258	896	21	34	+ 0.50	0.51
SX 20	19	14	238	278	37	48	− 0.85	1.20
SS 21	20	15	237	258	32	48	+ 0.70	0.88
SZ 40	12	16	230	159	12	48	+ 0.19	0.18
NT 36	31	17	202	416	27	30	+ 0.29	0.35

TR 61	13	18	195	184	25	35	+ 1.38	0.87
TG 63	11	19	180	142	20	28	− 4.05	1.03
NO 37	24	20	156	258	18	46	+ 1.02	0.58
NJ 38	19	21	153	190	30	51	− 0.26	0.51
TM 62	26	22	145	255	22	29	− 1.17	1.04
SM 12	4	23	143	46	43	45	+ 2.99	0.13
NH 28	8	24	131	70	29	46	− 0.66	0.29
NY 35	17	25	130	138	30	47	− 1.06	1.05
NR 16	2	26	117	14	24	29	− 13.39	0.76
TF 53	26	27	116	213	21	40	+ 0.15	0.18
SY 30	9	28	115	90	19	41	+ 1.58	1.33
SW 10	10	29	113	92	47	82	+ 1.03	1.47
NX 25	11	30	82	82	22	45	+ 0.09	0.06
SN 22	18	31	84	111	30	49	− 0.04	0.03
NN 27	6	32	79	33	32	39	+ 27.02	2.89
NC 29	3	33	68	10	29	37	+ 3.81	0.75
SH 23	24	34	56	89	24	50	− 1.13	0.61
NU 46	5	35	49	24	41	48	− 3.66	0.39
NM 17	2	36	24	8	18	34	− 4.27	0.33
Total	907	Total	378	22 301	12	82	+ 0.23	1.83

[a] Excluding four regions where there was no choice of TGR dose rates at district level.
* $p < 0.05$ (one-tailed).

Table 7 Factors included in a regression analysis of cancer associations

Factors: Description	Units	Mean values[a]
Sex	M : F ratio	1.43
Cancer onset age	Months	74
Year of birth	Calendar years	1959
Place of birth	Easting:Northing	43:32
TGR dose rate	nGy/h	32.0
Prenatal X-rays	% x-rayed	8.9%
Pregnancy illnesses	Per child	0.4
Postnatal infections	Per child	1.2
Social class	I to V	III
Maternal age	In years	28.9
Sibship position	1st to 4th etc.	2.2
Cancer clusters[a]	Average size	7.5

[a] Based on matched controls of OSCC cases.
[b] See Table 2.

Cancer rates and TGR dose rates. Crude comparisons at national level

Uncontrolled comparisons between cancer rates and TGR dose levels at national level produced no evidence of any cancer effects from *in utero* exposures to background radiation. On the contrary, the lowest of eight levels

Table 8 Results of the regression analysis

Factors		β^a	SE	*t* value	
Sex		−0.0105	0.0171	− 0.61	
Cancer onset age		−0.0003	0.0002	− 1.43	
Year of birth		0.0060	0.0014	+ 4.40	***
Place of birth	Easting	0.0059	0.0012	+ 4.85	***
	Northing	−0.0010	0.0006	− 1.58	
TGR dose rate		0.0065	0.0013	+ 4.97	***
Prenatal X-rays		0.6424	0.2285	+ 2.81	**
Pregnancy illnesses		0.0918	0.0096	+ 9.52	****
Postnatal infections		0.0645	0.0051	+12.74	****
Social class		−0.0272	0.0097	− 2.82	**
Maternal age		−0.0074	0.0015	− 4.80	***
Sibship position		−0.0083	0.0059	− 1.41	
Cancer clusters		0.0245	0.0013	+19.16	****

[a] β = change in log relative risk (RR) per unit change of each factor.
* $p < 0.05$.
** $p < 0.01$.
*** $p < 0.001$.

Table 9 Factors included in a second regression analysis and results of this analysis

Factors	Co-factors	β^a	SE	t value
Sex	—	−0.0104	0.0173	−0.60
Cancer onset age	—	0.0004	0.0003	−1.43
Year of birth	—	0.0059	0.0014	+4.26***
Place of birth (EE)	—	0.0058	0.0012	+4.70***
Place of birth (NN)	—	0.0008	0.0007	−1.22
TGR dose rate	—	0.0001	0.0019	+0.08
Prenatal X-ray	—	0.6505	0.2295	+2.83**
Prenatal X-ray	Cancer onset age	0.0003	0.0006	+0.54
Prenatal X-ray	Cancer onset age (squared)	$-2.4826(10^{-5})$	$0.9629(10^{-5})$	−2.57**
Prenatal X-ray	Year of birth	−0.0082	0.0036	−2.29**
Pregnancy Illnesses	—	0.1334	0.0140	+9.55****
Postnatal infections	—	0.0708	0.0072	+9.87****
Social class	—	−0.4660	0.0140	−3.34***
Maternal age	—	−0.0073	0.0015	−4.75***
Sibship position	—	−0.0088	0.0059	−1.50
Cancer clusters	—	−0.0095	0.0077	−1.24
Cancer clusters	TGR Dose Rate	0.0009	0.0002	+4.70***
Cancer clusters	Pregnancy illnesses	−0.0059	0.0015	−4.05***
Cancer clusters	Postnatal infections	−0.0008	0.0007	-8d-1.12
Cancer clusters	Social class	0.0029	0.0015	+1.29*

[a] See footnote to Table 8.

of TGR dose rates (under 20 nGy/h) had the highest cancer rate, and the highest dose level (over 50 nGy/h) had the third lowest cancer rate (Table 3). Even division of the country into three sections (by east to west aggregations of the large grid squares in Figure 1), still left the most westerly regions with the highest dose level and the lowest cancer rate, and the most easterly regions with the lowest dose level and the highest cancer rate (Table 4).

Therefore there was clearly a need to repeat the comparisons between cancer rates and TGR dose rates in circumstances which restricted the effects of all regional factors with cancer associations. Since each of the 911 districts with OSCC and NRPB data was contained within one of the large (100 km) squares in Figure 1, this could be done by treating the 40 large squares as separate population centres or 'zones'.

Cancer rates and TGR dose rates. With control for regional differences

When each of the large squares in Figure 1 was treated as a geographically distinct entity or zone, it soon became apparent that the seemingly favourable effects of background radiation seen in Tables 3 and 4 were false impressions caused by some unidentified factors having geographically related cancer associations which more than counterbalanced any effects of background radiation. This is most clearly seen by comparing the analysis of cancer rates and TGR doses at national level (Table 3) with a similar analysis of children living in the 100 km square containing Greater London (Table 5). This zone (see TQ 51 in Figure 1) included 61 districts with TGR dose rates ranging from 21 to 35 nGy/h. When arranged in this order, the lowest dose level (represented by three districts with 50 cancer deaths) had the lowest cancer rate, and the highest dose level (represented by one district with 40 cancer deaths) had the highest cancer rate. Furthermore, for the complete set of 15 dose levels, there was a significant upward slope of dose response (i.e. a regression coefficient of +1.30 with a t value of 4.01).

The large squares in Figure 1 included four coastal zones (with a total of 50 cancer deaths) where, at district level, there was no choice of TGR dose rates. However, for 20 of the remaining 36 zones there was an upward slope of dose response, including five zones in which this achieved statistical significance (Table 6). Among the remaining zones (with downward slopes of dose response) there were no significant regression coefficients. The zones with upward slopes of dose response accounted for 65 per cent of the cancer deaths and, after appropriate weighting, there was an overall upward slope of dose response (with a significant regression coefficient) for the whole country with 22 301 cancer deaths. Like the finding for cancer clusters (Table 2) this result was largely accounted for by densely populated districts. Thus TQ 51, SJ 33, NZ 45 and NS 26 not only accounted for 42 per cent of the cancer deaths, but also included three of the five zones with a significant regression coefficient.

Results of the regression analysis

With one exception all the mean values for the factors included in the regression analysis were based on matched controls of OSCC cases (Table 7). Therefore, they should be regarded as 'expected' values for the 9943 regional cohorts with cancer deaths. The single exception to this rule (see clusters in Table 7) was included in the regression analysis to discover whether the presence of one cancer in a cohort had 'attracted' other cases.

How many of the numerous 1944–79 illnesses of children and pregnant women in Britain had caused either postnatal deaths of premalignant infants or *in utero* deaths of premalignant fetuses was, of course, unknown. However, if the epidemics of infectious diseases (e.g. measles or influenza) had contributed towards these latency deaths, then the remaining cases (e.g. recognized cancers deaths) would certainly have a clustered distribution. Therefore, OSCC data for prenatal illnesses, postnatal infections and social class were included in the regression analyses both as factors with possible cancer associations (Table 8) and as factors with possible cluster associations (Table 9).

Other factors with known or suspected cancer associations were sex, birth-place, maternal age, sibship position, TGR dose rates and prenatal x-rays. Finally TGR dose rates were tested for cluster associations as well as cancer associations, and interactions between prenatal x-rays and two related factors (i.e. birth years and cancer onset ages) were also included in Table 9.

According to Table 8, factors with significant cancer associations included date and place of birth, TGR dose rates and prenatal x-rays, prenatal and postnatal illnesses, social class and maternal age. For most of these factors the introduction of cluster interactions and x-ray interactions into the regression analysis made no difference to the cancer associations. However, judging by the interaction effects, cohorts which had experienced an excessive number of cancer deaths (clusters) were largely responsible for the association between TGR dose rates and cancer rates (Table 9). Furthermore, for three factors (social class, pregnancy illnesses and postnatal infections) the associations with early cancer deaths were in one direction and the associations with clusters were in the opposite direction, and the clusters were clearly the result of interactions with pregnancy illnesses, postnatal infections, social class and TGR dose rates. Finally, judging by the x-ray interactions with date of birth and age at death, the obstetric x-ray cancer risk was much higher in the 1940s than in the 1970s and mainly affects cancer deaths between four and eight years of age.

DISCUSSION

According to the regression analysis we would expect a fetal dose of 1 nGy/h to add 0.0065 to the normal risk of experiencing a cancer death before 16 years of age (Table 8). Therefore, the doubling dose for **in utero** exposures to the gamma-ray component of background radiation could be in the region of 0.14 rem per annum. Alternatively from Table 6 we have a regression coefficient of 0.23 cases/million person years/nGy/h and an average cancer rate 22351/347 564 836 from Table 3 giving a doubling dose of 281 nGy/h or 0.25 rem per annum. According to various NRPB estimates the TGR dose rate averaged 32 nGy/h, and represented between 20 and 27 per cent of the total gamma dose.[8] Therefore the proportion of early cancer deaths caused by fetal exposures to gamma-radiation could lie between 66 and 96 per cent. Alternatively—on the assumption that 0.16 rem per annum is a typical background dose—we would expect fetal exposures to gamma-radiation to account for at least three-quarters of all juvenile neoplasms.

For prenatal x-rays—which typically require a third trimester exposure to half a rem—the logarithm of the relative risk was 0.6424 (Table 8). Therefore, the relative risk was 1.90. However, only 14 per cent of OSCC cases were x-rayed before birth. Therefore the proportion of OSCC cases attributable to this source of fetal irradiation probably lies between 6 and 7 per cent.

Other results included evidence of a non-random distribution of the OSCC cases which had left some regions and birth cohorts with an excessive number of deaths (cancer clusters) and some with exceptionally wide gaps between consecutive cases. Factors with cluster associations included prenatal and postnatal illnesses, social class, background radiation and population density, and for three of these factors (i.e. the prenatal and postnatal illnesses and social class) the cluster associations were in the opposite direction from the cancer associations. For example, the findings for social class 1 were indicative of a high cancer risk and a low risk of being part of a cancer cluster; and for non-fatal illnesses (before or after birth) the findings were essentially the same.

The findings for TGR dose rates were indicative of a cancer risk from background radiation, but this was only clearly seen in large (urban) cohorts with long follow-up periods. Therefore, since these cohorts accounted for most of the cancer clusters, the TGR dose rate features as one of several factors exerting independent cancer effects in Table 8, and as a cluster-related factor (with no independent cancer effects) in Table 9.

These observations lead to two conclusion: first, in relation to OSCC data the best chance of observing cancer effects of fetal exposure to background radiation was in densely populated regions where there was a wide range of TGR dose rates and continuous screening of cancer deaths from 0 to 16 years of age. Secondly, although we would expect leakages of radioactivity from nuclear power stations to produce genuine clusters of childhood leukaemia and other juvenile neoplasms, proof of such effects will require more evidence than has so far been produced .[9,10] For example, a basic requirement would be an epidemiological study capable of showing that (1) in the vicinity of *several* reprocessing plants or nuclear power stations there had been significant additions to TGR dose rates and (2) following this contamination (and after suitable intervals to allow for cancer latency) there was a significantly higher incidence of juvenile neoplasms than in a control group consisting of equally rural areas with comparable records of radiation dose rates and cancer rates.

Therefore, although the existence of leukaemia clusters in the vicinity of Sellafield and Dounreay is exactly what one would expect from this study, it is too soon to be quite certain that these clusters were a direct consequence of leakages of radioactivity from these reprocessing plants. Meanwhile, the most likely cause of clusters in other parts of Britain is cancer-induced loss of immunological competence leading to a clustered or epidemic distribution of latency deaths of preleukaemic children. Evidence of this type of interaction between measles and childhood leukaemias has been observed.[11] Therefore, it is possible that, in relation to juvenile neoplasms, both pregnancy illnesses and postnatal infections should be regarded as competing causes of death whose effects include a necessarily clustered distribution of latency deaths and recognized cancers.

APPENDIX 1

The statistical tests of cancer associations included uncontrolled and controlled comparisons between cancers and radiation doses and a special regression analysis. The latter incorporated Miettinen and Breslow[12,13] techniques for studying large series of cases with matched controls and proceeded as follows:

(1) Let $\mathbf{x}_i$ be a vector of observed values for case i, and let $\mathbf{y}_i$ be the vector of corresponding values for control i in a situation where $\mathbf{x}_i$ equals $\mathbf{y}_i$ for all matching or case-only factors.
(2) Let β be a vector of regression coefficients and α a constant such that the estimated probability of case i being an actual case $(p_i) = \exp(\alpha + \beta\mathbf{x}_i)$ and the corresponding probability for control $i(q_i) = \exp(\alpha + \beta\mathbf{y}_i)$.

In these circumstances α will equal the base-line logarithm of case incidence. Then, according to Miettinen (and with the stated values of $\mathbf{x}_i$ and $\mathbf{y}_i$), the conditional probability of any case/control pair having the values actually observed is $p_i/(p_i + q_i)$; and, according to Breslow, the best estimates of β are given by maximizing the log-likelihood

$$L_1 = -\sum_i \ln(1 + \exp[\beta(y_i - \mathbf{x}_i)])$$

Application to OSCC Data

(1) Let there be N_r cases for regional birth cohort r, and let the total person years of follow-up for r be T_r. Then the expected incidence for r is $Q_r = (\Sigma_{i \in r} q_i)/N_r$, where $\Sigma_{i \in r}$ means summation over N_r case/control pairs for regional birth cohort r.

In practice it is values of matched controls rather than the values of cancer cases which feature in this formula since the controls should be a random sample of the regional birth cohort and therefore more typical of regional birth cohorts than cases.

The expected value of N_r (M_r) is $T_r\, Q_r$, and the variance of N_r about this value is proportional to M_r. Therefore, best estimates of β and α are obtained by minimizing any discrepancy between N_r and M_r and this discrepancy $Z_1 = \Sigma_r (N_r - M_r)^2/M_r$.

This expression is similar to a chi-square formula, whereas the expression for L (above) is a log-likelihood. Therefore, as an alternative measure of discrepancy we have $Z_2 = \Sigma_r[N_r \ln(M_r) - M_r]$ since this is the variable part of the log-likelihood ratio statistic corresponding to Z_1 as a chi-square statistic.

As an expression to be evaluated Z_2 has two disadvantages: in the first place Z_2 depends upon control values. Therefore, it cannot be evaluated for regional birth cohorts with no cancer cases (and therefore no matched controls).

Secondly, although L_1 can be expressed as a sum over all case/control pairs this is not possible with Z_2.

The first problem was solved by treating N_r (and its expectation M_r) not as an ordinary Poisson variable (as in the conventional chi-square statistic derivation by Fisher) but as a zero suppressed Poisson variable. This meant adding an additional term to Z_2 and thus converting it to

$$Z_3 = \sum_r \left\{ N_r \ln(M_r) - M_r - \ln(1 - \exp[-M_r]) \right\}$$

The second problem was solved by realizing that when $\beta \mathbf{y}_i$ is small compared with α, any function $f(M_R)$ of M_R will be approximately equal to $\Sigma_{i \in r}$, $[f(N_r m_i)]/N_r$, where $m_i = T_r\, q_i/N_r$ is the contribution of case i to M_r.

Thus, we finally have an approximate log-likelihood $L_2 = -\Sigma_i\{m_i - \ln(m_i) + \ln(1 - \exp[-N_r m_i])/N_r\}$ which may be added to L_1 and the whole maximized for variation in α and β.

Finally, the statistical analysis recognized that there might be missing data, thus making it necessary to give median values for certain items such as sibship position (second), maternal age (25–29 years) and social class (grade III).

APPENDIX 2

Unsolved queries relating to the method of analysis in Appendix 1

We would like to point out that there are several unsolved queries requiring resolution before the method of Appendix 1 can be described as definitive.

The first query is as follows.

In the first approximation (Z_1) there are two sources of variance. The first is the Poisson variation of N_r about M_r; this is proportional to M_r, so there is variation of M_r about its true value. This variation comes about because there are only N_r controls to estimate the true mean value of Q_r. Fortunately, this extra variation is proportional to N_r, and may therefore be taken as proportional to M_r, when N_r is allowed to vary. Thus the total deviance of Z_1 or Z_2 will be multiplied by a constant factor.

This constant can be estimated as follows.

It can be shown that L_2 achieves its maximum value, with independent variation of the m_i when m_i is equal to e_i, with $e_i = 1 - \exp(-N_r e_i)$. This maximum value is given by $C = -\Sigma_i\, e_i - (1 - 1/N_r)\ln(e_i)$. Thus the deviance can be estimated by $V = -2\,(L_2 - C)$. A possible query is the number of degrees of freedom of this estimate. However, on a typical analysis with 21 factors, 22 351 cancers and 9943 cohorts with at least one cancer deaths, the value of V was 6221.84. Therefore, the deviance is probably less than any likely number of degrees of freedom, and equal weighting of L_1 and L_2 would seem to be appropriate.

The second query concerns the optimality of any linear combination of L_1 and L_2 in estimating α and β. Strictly speaking, this could only be justified if L_1 and L_2 were independent. However, considerations of orthogonality suggest that, since L_1 (the Miettinen–Breslow log-likelihood) is strictly a function of the difference between case and control values, the corresponding function of L_2 ought to be strictly a function of the sum of case and control values, rather than just control values only.

The third query concerns the statistical propriety of using (as one of the x's or independent factors in the regression) the value of N_r when this is clearly related to, if not itself the dependent factor. The context in which such a procedure suggests itself is as follows. Suppose, that an infection (in pregnancy or early childhood) either increases the chance of an early cancer death; or, alternatively, reduces the chance of detecting the malignant disease by causing a latency death. In such circumstances there would be a tendency for 'seed' cases to attract or repel subsequent 'descendant' cases, thus making it legitimate to regress the probability of any descendant case on the number of seed cases in a given cohort. In short the question reduces to, whether the overall cancer rate in a cohort can be legitimately regressed on the number of cancer cases when there is no means of distinguishing between seed and descendant cases.

ACKNOWLEDGEMENTS

This study was supported by grants from the Radiation Protection Programme, Contract No. B10-F-541-83-UK, of the Directorate-General for Research, Science and Education, Commission of the European Communities and the C. S. Fund, Freestone, California, USA.

REFERENCES

1. A. M. Stewart and G. W. Kneale, The age distributions of cancers caused by obstetric X-rays and their relevance to cancer latent periods. *Lancet*, **i**, 1185 (1970).
2. G. W. Kneale and A. M. Stewart, M-H analysis of Oxford Data: II. Independent effects of fetal irradiation subfactors. *J. Nat. Cancer Inst.*, **57**, 1009–14 (1976).
3. G. W. Kneale and A. M. Stewart, Age variation in the cancer risks from foetal irradiation. *Brit. J. Cancer*, **35**, 501–10 (1979).
4. A. M. Stewart, Childhood cancers and the immune system. *Cancer Immunol. & Immunother.*, **9**, 11–14 (1980).
5. G. W. Kneale and A. M. Stewart, pre-cancers and liability to other diseases. *Brit. J. Cancer*, **37**, 448–57 (1978).
6. A. M. Stewart and G. W. Kneale, The immune system and cancers of foetal origin. *Cancer Immunol. Immunother.*, **14**, 100–16 (1982).
7. G. W. Kneale, The excess sensitivity of pre-leukaemics to pneumonia. A model situation for studying the interaction of an infectious disease with cancer. *Brit. J. Prev. Soc. Med.*, **25**, 152–59 (1971).

8. J. S. Hughes and G. C. Roberts, The radiation exposure of the U.K. population—1984 review. *NRPB-R173*. HMSO, 1984.
9. *Investigation of the Possible Increased Incidence of Cancer in West Cumbria*. Black Report. HMSO, 1984.
10. M. A. Heasman, I. W. Kemp, J. D. Urquhart and R. Black, Childhood leukaemia in Northern Scotland. *Lancet*, **i**, 266 (1986).
11. A. M. Stewart, Epidemiology of acute (and chronic) leukaemias, in *Clinics in Haematology*, ed. S. Roach, p3. W. B. Saunders, London, 1973
12. O. S. Miettinen, Estimation of relative risk for individually matched series. *Biometrics*, **26**, 75–86(1970).
13. N. E. Breslow, N. E. Day, K. T. Halvorsen, R. L. Prentice and C. Sabia, Estimation of multiple risk function in matched case/control studies. *Am. J. Epidemiol.*, **108**, 209–307 (1978).

Radiation and Health
Edited by R. Russell Jones and R. Southwood

17

Childhood Leukaemia, Fallout and Radiation Doses near Dounreay

SARAH C. DARBY Ph.D AND RICHARD DOLL,* DM, FRS
Imperial Cancer Research Fund, Cancer Epidemiology and Clinical Trials Unit, University of Oxford, Radcliffe Infirmary, Oxford, UK

SUMMARY

The possible explanations of the recently reported increase in the incidence of childhood leukaemia around Dounreay are examined in the light of the changes in national leukaemia incidence that occurred during the period of exposure to fallout from international atmospheric testing of nuclear weapons. It is concluded that the increase cannot be due to underestimation of the risk of leukaemia per unit dose of radiation, nor to an underestimate of the relative biological efficiency of high as compared with low LET radiation. Possible explanations of the increase include an underestimate of the red bone marrow doses due to the Dounreay discharges relative to those from fallout, a misconception of the site of origin of childhood leukaemia, epidemics of infectious disease and exposure to some other unidentified environmental agent.

INTRODUCTION

Evidence has recently been published of an increased incidence of childhood leukaemia in the vicinity of two nuclear installations in the UK; namely, those at Sellafield,[1] where spent fuel from nuclear power stations is reprocessed, and at Dounreay,[2] where a prototype fast breeder reactor has been developed and where there is also a reprocessing operation that deals with fuel from a variety of reactors.

The National Radiological Protection Board (NRPB) has published calculations of the estimated radiation doses due to discharges from Sellafield and the

* Epidemiological consultant to the National Radiological Protection Board.

risk of leukaemia that might be expected in children living in the nearby village of Seascale.[3,4] Similar calculations have also been published for Thurso, the town nearest Dounreay.[5] These calculations are based on assumptions that have generally been accepted by radiobiologists and those concerned with radiological protection, and they lead to the conclusion that the doses received by the children are insufficient to have caused anything like the increase in leukaemia that has been observed in the vicinity of either installation. It is disturbing, however, that an increased incidence should have been observed near two separate installations and, especially, near the only two in Britain where a substantial amount of reprocessing of spent fuel takes place, and this demands further investigation. A fuller description of this work has been presented elsewhere.[6]

RADIATION EXPOSURES TO CHILDREN IN THURSO

Children living in Thurso are exposed to radiation from a variety of sources, the most important of which are: radiation from naturally occurring sources, fallout from weapons testing, medical x-rays, the discharges to sea and to atmosphere from Dounreay (throughout this chapter discharges to sea from Dounreay include those from the adjoining reactor site (Vulcan Naval Reactor Test Establishment) operated by the Ministry of Defence[7,8]), and the discharges to sea from Sellafield that have spread northwards round the coast. There is clear evidence that leukaemia can be caused by exposure to radiation and it is widely assumed that even very low doses will cause a slight increase in the risk of developing the disease. If this conclusion is true, then all the sources of radiation listed above will contribute to the risk of radiation-induced leukaemia experienced by children in Thurso.

For the comparison of the effects produced by different sources of radiation with those possibly produced by the discharges fallout has a number of advantages. The three recognized pathways of exposure for the discharges are: (i) external exposure from radionuclides deposited on the beach or on soil, or from the radioactive cloud, (ii) internal exposure from the inhalation of radionuclides in air or in resuspended beach sands and sea-spray, and (iii) internal exposure from ingestion of radionuclides in marine foods, especially shellfish, and in terrestrial foods, and the inadvertent ingestion of beach sands and soil. All three of these exposure pathways are also thought to operate for fallout.

Fallout, moreover, has a number of radionuclides in common with the discharges. Many of these are capable of irradiating the red bone marrow which, it is assumed, is the tissue in which leukaemia is induced. Those that contributed to the dose to the red marrow in children born in Thurso in 1960 are listed by Dionian *et al.*[5] in two groups according to the type of radiation produced (low or high LET) for each source and are reproduced in Table 1.

Table 1 Radionuclides occurring in fallout, or in discharges from Dounreay or Sellafield and contributing to the radiation dose to the red bone marrow of a child born in Thurso in 1960. (Data derived from Dionian *et al.*[5])

Sources of Radiation	Radionuclides	
	Low LET (includes x-rays, gamma-rays)	High LET (includes alpha particles)
1. Fallout, Dounreay discharges[a] and Sellafield discharges[b]	^{90}Sr, ^{106}Ru, ^{137}Cs, ^{144}Ce, ^{239}Pu, ^{238}Pu,[c] External radiation	^{239}Pu, ^{238}Pu[c]
2. Fallout, Dounreay discharges[a]	^{89}Sr, ^{131}I	—
3. Dounreay[a] and Sellafield[b] discharges only	^{95}Zr, ^{95}Nb, ^{134}Cs, ^{241}Pu,[d] ^{241}Am[d]	^{241}Pu,[d] ^{241}Am[d]
4. Fallout only	^{14}C[e]	—

[a] Discharges to sea and to atmosphere.
[b] Discharges to sea.
[c] ^{238}Pu is not listed explicitly by Dionian *et al* as occurring in Dounreay and Sellafield discharges, but a small amount does occur and it has been taken into account by Dionian *et al*, in their calculations (J. Dionian, personal communication).
[d] ^{241}Pu, ^{241}Am are produced in fallout, but the contribution to dose is very small by comparison with other radionuclides.
[e] ^{14}C is present in Dounreay discharges, but only in small quantities compared with other radionuclides.

This contrasts sharply with the situation for medical and natural irradiation, the other two major sources of exposure which have no radionuclides in common with these discharges.

Lastly, fallout has the advantage for our present purpose that the resulting exposure to radiation has varied greatly in time.[9] The first atmospheric nuclear weapon test took place at Alamogordo in 1945, prior to the bombings of Hiroshima and Nagasaki. Atmospheric testing activities increased after the Second World War, and reached a peak in the late 1950s and early 1960s. Following the 1962 test ban treaty the level of atmospheric testing has remained low, with only France and China continuing to test.

Discharges from the Dounreay site commenced in 1958[8] and from Sellafield around 1950.[4] The average annual absorbed doses from low and high LET radiation received in the bone marrow by children born in Thurso from these two sources and from fallout have been estimated separately by Dionian *et al.* Detailed figures have been published for the cohort born in 1960. The contributions from Dounreay and fallout are shown graphically in Figure 1; the maximum contributions from Sellafield for both high and low LET radiation are small by comparison with the maximum contributions from either fallout or Dounreay and are omitted.

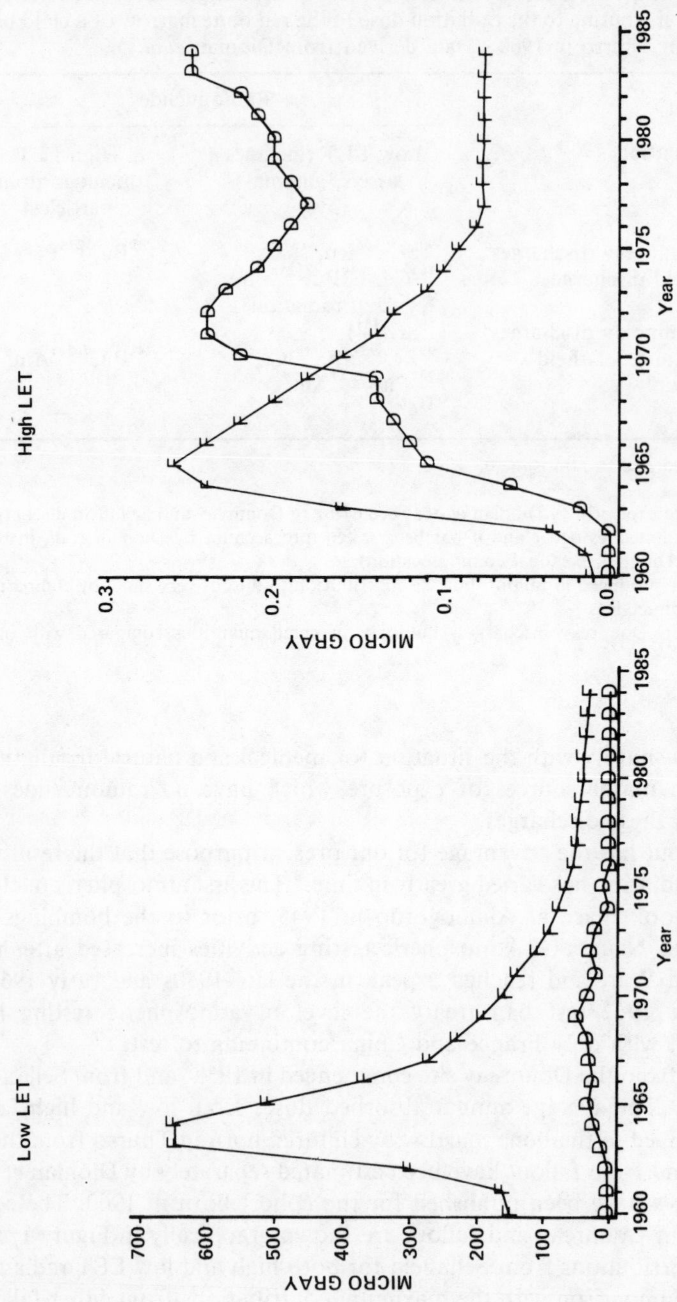

Figure 1 Estimated average annual absorbed radiation doses to the red bone marrow resulting from fallout (F) and Dounreay discharges (D) for high and low LET radiation for an individual born in 1960 in Thurso, by year. Fetal doses are not included on the slide, but are estimated to be low LET: fallout 86 μGy, Dounreay 8.4 μGy; high LET: fallout 6.4×10^{-4} μGy, Dounreay 7.3×10^{-5} μGy. (Data derived from Dionian *et al.*[5])

From Figure 1, it is seen that the estimated annual dose from low LET radiation due to fallout rose to a maximum a few years after the period of peak testing activity and then declined. In every year, however, the estimated dose due to fallout was greater (and sometimes much greater) than that due to the Dounreay discharges. The estimated dose from high LET fallout also reached a peak in the mid-1960s and then declined. By the late 1970s the decline had ceased and the estimated annual dose subsequently remained approximately constant. This constant dose is due to inhaled ^{239}Pu, which is thought to have a long retention time on bone surfaces, and to irradiate the red bone marrow continuously. The estimated annual high LET dose from the Dounreay discharges tended to increase throughout the time period and from about 1970 it exceeded that from fallout. The maximum estimated annual high LET dose due to Dounreay discharges occurred in the early 1980s and is comparable with the estimated dose due to fallout in the mid-1960s.

POSSIBLE EXPLANATIONS OF THE OBSERVED INCREASE IN LEUKAEMIA INCIDENCE IN THE VICINITY OF DOUNREAY

Four explanations for the observed increase in leukaemia incidence in the vicinity of Dounreay seem possible. These are summarized in Table 2.

1. Some leukaemias caused by discharges, and some leukaemias also caused by fallout

One explanation is that most of the leukaemias near Dounreay are, in fact, caused directly by the discharges. If this is true, it follows that doses from fallout must also have been large enough to cause a material increase in the risk of childhood leukaemia, unless either explanation 2 or explanation 3 (see Table 2) is also true. Fallout, however, is not specific to the area around Dounreay, and this enables the effect of fallout to be tested.

If the doses from fallout were large enough to cause a material increase in childhood cancer rates, it would be expected that the effect would be clearly visible in cancer incidence data from Denmark and Norway, where national

Table 2 Summary of possible explanations of observed increase in leukaemia incidence in the vicinity of Dounreay

1. Some Dounreay leukaemias caused by discharges, and doses from fallout also large enough to increase childhood leukaemia incidence materially.
2. Estimates of relative magnitude of average red marrow doses due to fallout and discharges wrong.
3. Red bone marrow dose irrelevant for childhood leukaemia.
4. Dounreay leukaemias caused by some agent other than radiation exposure from discharges.

cancer registries have long been established and where the temporal trend in fallout was the same as in Britain. For high LET radiation the estimated doses in these two countries are similar to those in Britain. For low LET radiation the estimated doses are higher than in the UK especially in Norway, because of its greater level of rainfall. Measured levels of ^{137}Cs in Norwegian milk have been about six times those in the UK, and measured levels of ^{90}Sr about three times the UK values, while the levels in Denmark have been intermediate between those for Norway and the UK for ^{90}Sr and similar to the UK values for ^{137}Cs.[9]

Data on the incidence of acute leukaemia (which would include almost all childhood leukaemia) are available from 1943 for Denmark and from 1953 for Norway.[10] These are shown in Figure 2 standardized for age. In neither country is there a peak in incidence in the late 1960s or in the early 1970s following the peak in doses from fallout. The increasing trend in Denmark prior to 1960 is thought to have been at least partly artificial and due to an increase in the proportion of children with leukaemia that were correctly diagnosed.[12] Data on childhood leukaemia mortality and registrations in Britain confirm this conclusion.[6]

Although the rates of childhood leukaemia have varied in each of these three countries, there is no convincing evidence of an increase in incidence which could be attributed to fallout, leave alone any comparable with that observed around Dounreay.

The arguments presented in this chapter make the conventional assumption that risk per unit dose is approximately uniform throughout the range of doses that have been received by children living in Thurso. It has, however, sometimes been suggested that for low LET radiation the risk per unit dose varies within the low dose range, resulting in a higher risk per rad for those at the upper end of the range, and making the average dose an inappropriate measure. However, the fact that the average annual doses from weapons fallout have varied up to levels that are well over ten times the maximum annual average discharge dose, means that if this were so an even greater effect of fallout in comparison with the discharges would be expected. For high LET radiation, radiobiological work does not, on the whole, suggest a departure from a uniform risk per rad, even in the low dose range.

2. Estimates of the relative magnitude of average red marrow doses due to fallout and discharges wrong

A second possible explanation of the findings is that the true red marrow dose due to the discharges when averaged over all individuals is much higher than that due to fallout. A popular explanation postulates the existence of a small subgroup of individuals who have received very high doses due to the discharges and who have been omitted from the calculations. We have

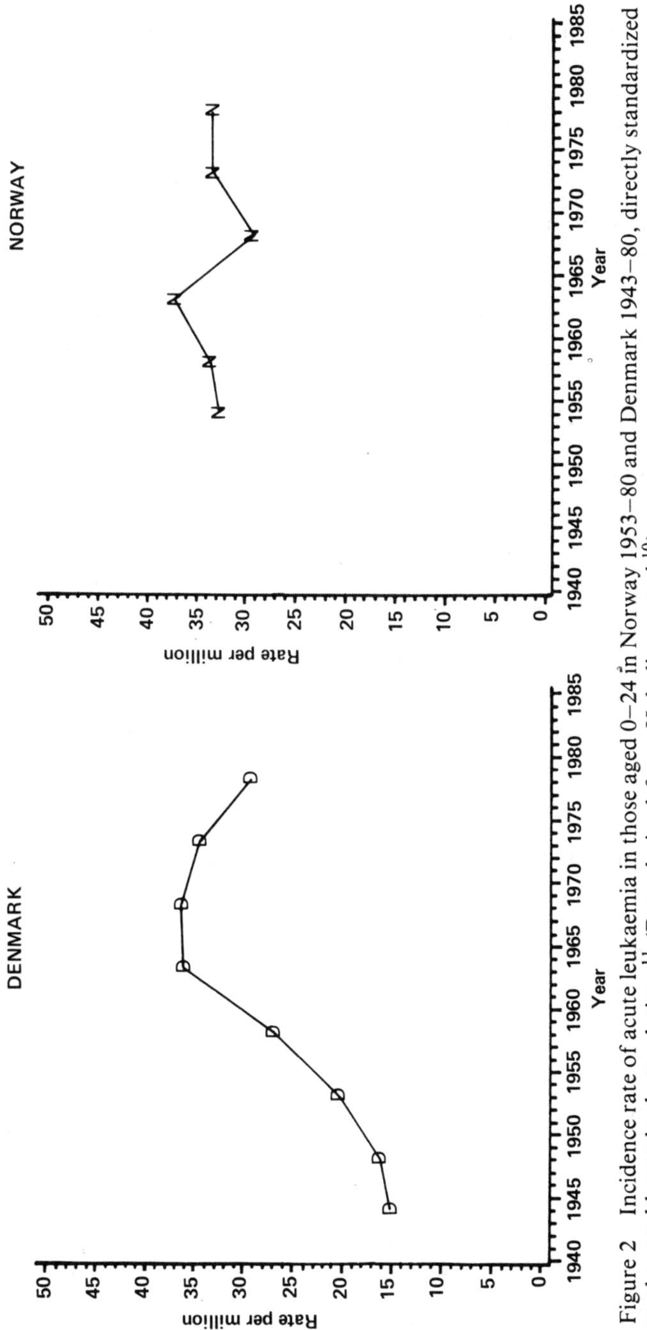

Figure 2 Incidence rate of acute leukaemia in those aged 0–24 in Norway 1953–80 and Denmark 1943–80, directly standardized to the world standard population.[11] (Data derived from Hakulinen *et al.*[10])

discounted this as accounting for the findings for the following reasons. Firstly, even if a very small proportion of the children had received very high doses and had not been taken into account in the calculations, this would have a negligible effect on the overall average dose; for example, if 1 per cent of children had in fact received ten times the estimated average dose and had been omitted then altering the calculations to take them into account would increase the average by only 10 per cent. Thus, to explain the findings, it is necessary to postulate that the proportion of children receiving extraordinarily high doses is substantial. The mechanisms by which this might occur, for example, living on the beach, eating a diet of pure shellfish, or deliberate ingestion of large quantities of beach sands, are such that it is not plausible that they apply to a substantial proportion of children. There are, however, a number of ways in which it might theoretically be possible for a large discrepancy in the estimates of the relative magnitude of doses from fallout and from the discharges to have occurred.

First, the levels of radioactive materials assumed to be present in the environment might be wrong. Those from fallout are based on environmental measurements and are, therefore, unlikely to be grossly in error. Those from the discharges are based primarily on mathematical models.[8] The results obtained by comparing the model predictions with such environmental measurements as have been made suggested that the predicted doses are unlikely to be far out. Indeed the doses predicted by the models appear to overestimate the concentrations of radionuclides in the environmental materials that are thought to be the most important causes of exposure by a factor of two or three.[8]

Secondly, assumptions regarding individual behaviour such as the amount of time spent on the beach, the quantities of various different types of food eaten, etc., might be wrong; and thirdly, the metabolic models relating the dose in an individual organ or tissue to an intake of radionuclide might be wrong. As it is impossible to measure the body content of Pu *in vivo* at the levels of interest, it is not possible to test all the assumptions directly. However, in two studies carried out comparing measured and predicted body content of ^{137}Cs, the models have tended to overestimate dose.[13,14] In Germany, measurements of the concentration of Pu due to weapons fallout have also been made at autopsy for a number of organs and good agreement has been shown between measured and predicted values.[15] It should also be pointed out that the same metabolic models are used for both discharges and fallout and thus cannot distort the relative magnitude of the dose from the two sources in so far as the same radionuclides and methods of intake are involved. For high LET the majority of dose from both fallout and the discharges is due to ^{239}Pu. However, for fallout the dose is received mainly via inhalation, while for the discharges it is received mainly via ingestion. Substantial differences could, therefore, be produced if the assumed proportion of Pu transferred from the lung into the blood stream is much too high, or the assumed proportion transferred from the digestive tract is much too low.

3. Red bone marrow dose irrelevant for childhood leukaemia

The target cells for adult myeloid leukaemia are known to be located in the bone marrow, and according to Professor Melvyn Greaves, director of the Leukaemia Research Fund's centre at the Institute for Cancer Research, there is now evidence that a substantial proportion of childhood leukaemias also arise in the bone marrow, with a small proportion of cases arising in the fetal liver or the thymus. The question is, however, not completely resolved (M. Greaves, personal communication).

It is uncertain as to where in the bone marrow the target cells lie, but even if they are restricted to a small region of the marrow, it seems unlikely that the ratio of fallout dose to discharge dose for that region could differ substantially from that for the marrow as a whole.

4. Dounreay leukaemias caused by some agent other than radiations exposure from discharges

The fourth possible explanation is that the Dounreay leukaemias are not caused by radiation exposure from the discharges, but by some other environmental agent. It is not easy, however, to suggest what such an agent might be. The only agents, other than ionizing radiations, that are known to have caused leukaemia are benzene, some of the alkylating agents and other drugs used for the treatment of cancer, and two viruses (the Human T-cell Lymphotropic Viruses types 1 and 2) and none of these have caused the typical acute lymphoblastic leukaemia of childhood.

Viral infection has often been suspected as the most likely principal cause; but no direct evidence of such a cause has been obtained. Reports of clusters of cases have recurred frequently in the last 50 years but the evidence that clusters may occur more often than would be expected by chance alone is not conclusive.[16] If an infective agent is responsible for the disease, it is possible that mini-epidemics would be particularly likely to occur under the conditions that have brought together a young workforce in a relatively isolated area.

Many other environmental agents have been suspected as possible causes from time to time (for example, electromagnetic fields,[17] parental occupation particularly paternal exposure to chlorinated solvents, pesticides and incense (J. Peters, personal communication)). None has yet been substantiated, and there is no obvious reason, other than parental occupation in a nuclear installation, why such factors should have caused an increased number of cases specifically round Dounreay and Sellafield.

CONCLUSIONS

The available data weigh heavily against the idea that the recent increase in childhood cancer near Dounreay could be accounted for by radiation exposure due to the discharges, unless the doses to the red marrow from the discharges are grossly underestimated by comparison with the doses received from fallout. In particular, the possibility that an underestimate of either the risk of leukaemia per unit absorbed dose to the bone marrow at low doses and low dose rates, or of the relative biological efficiency of high LET as compared with low LET radiation is ruled out by the lack of any consistent increase in childhood leukaemia following the period of peak fallout. Investigation of the various ways in which underestimates of dose might have occurred does not suggest any ready explanation but there remains the possibility that childhood leukaemia does not arise from the red marrow and that the radioactive nuclides are disproportionately concentrated in the relevant tissue. Alternatively the increase might be the result of a cluster of cases brought out by an infective agent or perhaps by some other environmental agent. Which, if any, of these explanations is correct can be determined only by further research.

ACKNOWLEDGEMENTS

Thanks are due to Mrs I. Stratton for computing assistance, Mrs V. Weare for clerical assistance, and Mrs C. Harwood for preparing the manuscript.

This chapter is an abbreviated account of the paper given at the conference. Full details have been published in the *British Medical Journal* of 7 March 1987 and we are grateful to the Journal for allowing much of that article to be reproduced here.

REFERENCES

1. M. J. Gardner and P. D. Winter. Mortality in Cumberland during 1959–79 with reference to cancer in young people around Windscale. *Lancet*, 216–17 (1984).
2. M. A. Heasman, I. W. Kemp, J. D. Urquhart and R. Black, Childhood leukaemia in Northern Scotland. *Lancet*, i 266 and 385 (1986).
3. J. W. Stather A. D. Wrixon and J. R. Simmonds, The risk of leukaemia and other cancers in Seascale from radiation exposure. *NRPB Report R171*.London HMSO, 1984.
4. J. W. Stather, J. Dionian, J. Brown, T. P. Fell and C. R. Muirhead, The risks of leukaemia and other cancers in Seascale from radiation exposure. Addendum to NRPB Report R171. London: HMSO, 1986.
5. J. Dionian, C. R. Muirhead, S. L. Wan, and A. D. Wrixon, The risks of leukaemia and other cancers in Thurso from radiation exposure. *NRPB Report R196*. London: HMSO, 1986.
6. S. C. Darby and R. Doll, Fallout, radiation doses near Dounreay and childhood leukaemia. *Br. med. J.*, **294**, 603–607 (1987).

7. G. J. Hunt, Aquatic Environment Monitoring Report, Lowestoft: Ministry of Agriculture, Fisheries and Food, 1986.
8. M. D. Hill and J. R. Cooper, Radiation doses to members of the population of Thurso. *NRPB Report R195*. London: HMSO, 1986.
9. United Nations Scientific Committee on the Effects of Atomic Radiation (UNSCEAR). *Ionizing Radiation: Sources and Biological Effects*. New York: United Nations, 1982.
10. T. Hakulinen, A. H. Anderson, B. Malker, E. Pukkala, G. Schon and H. Tulinius, Trends in cancer incidence in the nordic countries. *Acta Pathologica*, **94**, Suppl. 288, Section A (1986).
11. J. Waterhouse, C. Muir, K. Shanmugaratnam and J. Powell, *Cancer Incidence in Five Continents*. Vol. IV. Lyon: International Agency for Research on Cancer, 1982.
12. N. E. Hansen, H. Karle and O. M. Jensen. Trends in the incidence of leukaemia in Denmark, 1943–77. *J. Natl. Cancer Inst.*, **71**, 697–701 (1983).
13. F. A. Fry and T. J. Summerling, Measurements of Caesium-137 in Residents of Seascale and its Environs. *NRPB Report R172*. London: HMSO, 1984.
14. F. A. Fry, Comparison of measured and predicted body contents of ^{137}Cs in adults resident near Sellafield. Appendix D of Addendum to *NRPB Report R171*. London: HMSO, 1986.
15. K. Bunzl, K. Henrichs and W. Kracke, Distribution of fallout ^{241}Pu, $^{239+240}$Pu and ^{238}Pu in persons of different ages from the Federal Republic of Germany, in *Assessment of Radioactive Contamination in Man 1984*. Vienna: International Atomic Energy Agency, 1985.
16. P. G. Smith, Spatial and temporal clustering, in *Cancer Epidemiology and Prevention*, eds D, Schottenfeld and J. F. Fraumeni, Jr. Philadelphia, PA: Saunders, 1982.
17. N. Wertheimer and E. Leeper, Electrical wiring configurations and childhood cancer. *Am. J. Epid.*, **109**, 273–84 (1980).

8. [illegible] Radiation Monitoring Report, [illegible] Ministry of [illegible] 1962.
9. M. [illegible] and J. R. [illegible] Radiation doses [illegible] the population of [illegible] London: HMSO [illegible].
10. United [illegible] Scientific Committee on the Effects of Atomic Radiation [illegible] New York: United Nations, 1962.
11. J. [illegible] Pollution [illegible]
12. [illegible] Hansen [illegible] and [illegible] Trends in incidence of leukaemia in Denmark [illegible] Cancer [illegible]
13. F. A. Fry and [illegible] Strontium [illegible] Measurements [illegible] London: HMSO [illegible]
14. F. A. Fry [illegible] NRPB Report [illegible]
15. [illegible] and W. [illegible] Determinations of [illegible] in [illegible] of different ages from the Federal Republic of Germany [illegible] International [illegible] Agency [illegible]
16. D. Smith [illegible] Philadelphia: [illegible] Saunders, [illegible]
17. [illegible]

Radiation and Health
Edited by R. Russell Jones and R. Southwood

18

Leukaemia and Nuclear Power in Britain

JOHN URQUHART
Gosforth, Newcastle upon Tyne, UK

INTRODUCTION

Ever since the YTV programme, 'Windscale, the Nuclear Laundry' showed a tenfold excess of childhood leukaemia near the Sellafield nuclear reprocessing plant, there has been considerable public and scientific debate about the possible connection between nuclear power and leukaemia. In this chapter we look at the evidence published so far and give the results of an investigation into young leukaemia death rates around fourteen nuclear installations in England and Wales in the period 1963 to 1980.

Before the YTV programme was shown in November 1983 it was assumed that the amount of radioactivity discharged from nuclear plants could not cause a detectable number of extra cancers, although it was accepted that the amount of radioactivity discharged from the Sellafield (Windscale) nuclear reprocessing plant could cause increased cancer risk both locally and nationally. The discovery of five leukaemia cases under ten over a 25-year period in Seascale,[1] the village next to Sellafield, when normally only one case would be expected in sixty years aroused considerable public interest, and within 48 hours the British Government authorized a special investigation under Sir Douglas Black.

The first task of the Black committee was to confirm or dismiss the claims made in the YTV programme, and this was done by comparing the observed number of young leukaemias with the national rate. The committee also looked at young lymphomas, which are also RES neoplasms. They confirmed the YTV findings and furthermore showed that Seascale was the worst area for young lymphoid malignancies in the Northern region in recent years.[2]

Although the Black Report accepted that the observed number of leukaemias were highly unusual they argued that the amount of radiation released from Sellafield was not high enough to account for the number of extra leukaemias observed. Nevertheless, they set up further research into possible links between radiation and leukaemia.

STATISTICAL SIGNIFICANCE

Could the Sellafield findings be accounted for by chance? Statistical analysis gives no absolute proof of a cause and effect link, but when findings are unusually high, that is have less than one in twenty probability of occurring by chance, then it is generally accepted that further investigation should be made. The Sellafield leukaemias and lymphomas had less than 1 in 10 000 chance of occurring naturally.

After the YTV programme there were reports of leukaemia excesses or clusters all over the country—some near nuclear facilities, some not. Obviously some clusters of leukaemias up to a certain level can occur by chance. For example, if we were looking at say, 400 cases in 500 areas of equal population there would be about 25 areas where the leukaemia rate observed has a 1 in 20 chance or less of occurring and there would be one area whose level was so high that it would have a 1 in 500 chance of occurring. What if the observed rate in one area is even higher than might occur by chance? This could be called a super cluster. Such an observation would be outside the normal Poison distribution. Mathematically it can be defined as occurring when the probability of the observed event in a particular area is less than the inverse of the total number of areas examined. In the Black report, Doctors Craft and Openshaw compared the Seascale rate with the 674 other wards in the Northern Region.[3] Not only were the Seascale ward figures the least likely to occur by chance, but, in terms of unlikeliness were well above all other wards and, with a probability of only 1 in 10 000, fitted the definition of a super cluster. Craft and Openshaw did not extend their investigations to include the last two cases found in Seascale in 1983. If they had the probability would have been less than 1 in 100 000.[4]

SCOTTISH DATA

A more comprehensive investigation of leukaemia patterns were carried out by the Scottish Health Office. Dr Heasman and colleagues looked at over 900 postal sectors in Scotland, equivalent to wards. One postal sector fulfilled the definition of a super cluster. It was the sector that contained the Dounreay nuclear reprocessing plant.[5] The under 25 leukaemia rate was ten times the national average for the period 1979–84. Equally important was their finding that in Scotland as a whole the distribution of all other leukaemias fitted with a Poisson distribution and could therefore be accounted for by chance.[6]

Dr Heasman and his colleagues were originally asked to find out whether there were any unusual levels of leukaemia around nuclear installations in

Scotland. They tested this hypothesis by drawing circles of varying radius around the installations and seeing whether the observed cases of lymphoid and myeloid leukaemia around nuclear installations in Scotland were greater than expected.

Table 1 demonstrates that within a 12.5 km radius other installations apart from Dounreay show excess cases of childhood leukaemia. The increase is not as significant as for Dounreay (i.e. there are no other super clusters) but combining the data from all five nuclear facilities produces elevated incidence rates for all the time periods (SRR of 118 for 1968–73, 115 for 1974–8 and 136 for 1979–84).

Dr Heasman and colleagues extended their analysis to include childhood lymphoma (as well as leukaemia) within a 6.25, 12.5 and 25 km radius around the Scottish nuclear installations. Table 2 demonstrates that within 6.25 km of Rosyth Naval Dockyard there were thirteen cases from 1974–78 against an expectation of less than five. This again fulfils the definition of a super cluster when compared with expected rates in equivalent urban areas ($p < 0.0015$).

Needless to say publication of such leukaemia figures generates the counterclaim that clusters occur naturally at sites remote from nuclear facilities. For example, Dr Cartwright of the Yorkshire Cancer Registry analysed regional leukaemia patterns by postal sector for all ages over a six-month period.[7] He suggested a clustering effect as his results did not fit a Poisson distribution. His findings, however, were marred by the assumption that postal sections have equal populations.[8] A later paper by Dr Cartwright on variation of incidence failed to point out that the leukaemia rate in Cumbria was 100 per cent higher than comparable regions.[9] Nevertheless other environmental cases of

Table 1 Young leukaemias in Scotland

Area	1968–73		1974–8		1979–84	
	O	E	O	E	O	E
Chapel Cross	2	1.19	0	1.14	3	1.11
Dounreay	2	1.19	0	0.47	5	0.47[a]
Hunterston	3	3.35	5	3.18	6	3.11
Holy Loch	8	5.03	5	4.66	4	4.48
Rosyth	10	9.39	12	9.64	8	9.94
Total	23 *v* 19.44		22 *v* 19.09		26 *v* 19.11	
SRR	118		115		136	
Total	1968–83					
	O	E				
	71 *v* 57.64					

[a] Super cluster.
From *Leukaemia and Young Persons in Scotland*, SHS Evidence to the Dounreay Inquiry.

Table 2 Observed and expected cases of non-Hodgkin's lymphoma, lymphoid and myeloid leukaemia within 6.25 km of Holy Loch and Rosyth. (0–24 age group)

Area	1968–73		1974–8		1979–84	
	O	E	O	E	O	E
Holy Loch	1	0.89	2	0.82	2	0.89
Rosyth	7	5.30	13	[a]4.89	4	5.33
Total	8 *v*	6.19	15 *v*	5.71	6 *v*	6.22
SRR	129		203		96	
Total	O	E				
	29 *v*	18.12				

[a] $P < 0.0015$.

leukaemia cannot be ruled out. But it should be emphasised that discovery of a second cause of leukaemia would in no way invalidate any significant results around nuclear installations. After all, non-smokers die of lung cancer, but that does not mean that smoking cannot cause lung cancer, and if a second cause of leukaemia were found this would imply that the natural level of leukaemia was even lower, making any excess round nuclear installations even more significant.

STUDY OF FOURTEEN NUCLEAR INSTALLATIONS IN ENGLAND AND WALES

One way of assessing the problem further is to examine the incidence of childhood leukaemia around nuclear plants and compare them not only with national levels but similar areas free of nuclear influences. But there is a more fundamental philosophical reason for abandoning the cluster approach. It may be appropriate for investigation but not for analysis without a prior hypotheses, and the arbitrary collection of clusters without a prior hypothesis does not advance the scientific method. Nor is it helpful, for example, to observe that there is an increased rate round, say, ancient military fortifications since theoretically there is an infinite number of alternative hypothesis to choose from. There is a prior hypothesis to be tested on the table. Namely that there is an increased young leukaemia rate in the areas immediately surrounding nuclear installations.

METHODS

This study is based upon publicly available leukaemia mortality data in the OPCS records in London. Between 1963 and 1980 these are available for all rural and urban districts in England and Wales. These areas were abolished in 1974 but they produced a useful framework for leukaemia studies as they are smaller than the country districts that replaced them. Under-25 age group leukaemia death rates were compared with national deaths and rates in control areas.

All methods of choosing boundaries around nuclear plants have their problems particularly in England and Wales where there is no exact correspondence between local government districts such as wards and postal codes. Studies that use confidential information can be subject to error. For example, the initial paper by Craft and Openshaw lost vital information in West Cumbria because the computer program placed some cases in the Irish Sea![10,11] On the other hand, publicly available information has to follow local government boundaries, which are irregularly shaped. The present study used publicly available information based on local government areas that lay wholly or partly within 12.5 km of plants known to discharge radioactivity. The 12.5 km radius circle has already been used for analysis in Scotland. The plants in England and Wales consisted of eight nuclear power plants and six nuclear facilities or radiochemical facilities, excluding Sellafield.

The eighteen-year period between 1963 and 1980 was divided into three, six-year periods. The use of shorter periods is more likely to pick up any significant short-term effects while any overall effects can be compared between different time periods.

The leukaemia death rates round each nuclear installation were compared with control areas. These were chosen to include all those in urban and rural areas wholly or partly within a circle of 12.5 km radius away from the plant. The centre of control circles were chosen to be 50 km up-wind of the nuclear installation. However, where such circles lay down-wind of other nuclear plants, or off the coast or in areas that were not similar to the primary sites, these centres were placed up to 75 km from the plant. Some nuclear radioactivity can travel long distances—witness Chernobyl—so it was important to choose clean areas which should be free of possible nuclear influences. For example, for Sizewell the control area was chosen in a similar rural district 75 km north of the plant and not in the rural area north-west of the plant. This latter area lies down wind of the Lakenheath air-base where a nuclear accident occurred in 1957. It was one of the districts in that area, Depewade, that was reported by the Black committee to have the highest leukaemia rate in a rural district in the country, even higher than the leukaemia rate around Millom, the rural district south of Sellafield. No causal link between the high rate in Depewade and the Lakenheath fire has been proved but for the purpose of choosing a control area such a possibility has to be excluded.

The rural and urban districts in the chosen areas can also be compared in different ways. For example, proximity to plant, whether up-wind or down-wind and so on. It might also be expected that the amount of outdoor exposure and consumption of local foodstuffs would vary between town and country, and this might have a bearing on the pathways of radioactivity reaching local people. In addition, combining large population centres with small rural areas might swamp out detectable health variations. Results for urban and rural areas were therefore considered separately. The under 25 population in all the urban areas was about 600 000 and in the rural areas nearly half a million. Comparisons were made with national and control areas by using population data for the age groups 0 to 4, 5 to 14 and 15 to 24.

RESULTS

The results for eighteen years are set out in Table 3. It will be seen that there is an excess of young leukaemia in rural areas with a probability by chance of less than two in a thousand. Two individual sites were statistically significant—Hinkley Point ($P = 0.04$) and Springfields ($P = 0.01$). It must be remembered that this study does not include the leukaemia rates in the rural area round Sellafield, which have already been shown to be highly significant.

Table 3 Young leukaemia death rates around fourteen nuclear installations in England and Wales by rural and urban districts wholly or partly within 12.5 kilometres (0–24 age group) 1963–80

Installation	Urban		Rural		
	O	E	O	E	
Aldermaston and Burghfield	25	30.59	58	46.98	$P = 0.06$
Amersham	58	50.73	38	32.88	$P = 0.15$
Berkeley and Oldbury	2	1.41	25	22.55	
Bradwell	5	4.42	8	8.77	
Capenhurst	40	44.46	12	12.65	
Devonport	43	42.08	9	5.13	$P = 0.08$
Dungeness	1	1.13	1	0.96	
Harwell	8	5.89	11	14.01	
Hinkley Point	7	5.43	11	5.93	$P = 0.04$
Sizewell	1	0.92	7	4.08	$P = 0.12$
Springfields	28	35.78	32	20.66	$P = 0.01$
Trawsfynydd	1	1.25	1	0.88	
Winfrith	11	8.79	7	6.89	
Wylfa (1971–4)	0	0.27	2	0.93	
Totals	230	233.15	222	183.3	
	n.s.			$P = 0.002$	

Table 4 Comparisons of under-25 leukaemia death rates in rural, urban and rural control areas in six-year periods 1963–1980.

	1963–8		1969–74		1975–80		Totals	
	O	E	O	E	O	E	O	E
Urban	81	81.0	78	80.3	71	72.8	230	233.1
Rural	72	62.4	86	64.7	62	56.2	222	183.3
			$p = 0.01$					
Rural control	32	31.0	23	26.2	28	28.8	83	86

The young leukaemia rates in the rural districts around the nuclear installations are also significant against national rural rates. A study of national rural districts showed that young leukaemia rates were slightly higher than the national levels but this excess could almost be accounted for by the extra deaths in the rural areas next to the nuclear plants. The rural results were also compared with those in the rural districts in the control areas. Again they were highly significant as there was no difference between the rates in the control areas and national rates.

The rural results were also analysed in three six-year periods, 1963–8, 1969–74, 1975–80. It can be seen that the excess at Hinkley Point could be accounted for by the early years whereas the levels at Springfield remained high throughout the eighteen-year period. Of the fourteen nuclear installations, ten had above average young leukaemia rates in the rural districts surrounding them. A comparison of rural, urban and control rates is shown in Table 4.

DISCUSSION

The observed difference in young leukaemia rates between rural and urban communities near nuclear installations may account for some of the effects previously described. There is an increase in young leukaemia death rates in the rural areas surrounding nuclear installations in England and Wales which discharge radioactivity. Between 1969 and 1974 for example, this was more than 30 per cent above normal levels and for the year period was greater than 20 per cent.

There is a need to carry on investigating and recording cancer figures for deaths and cases since 1974. Already local groups have identified further leukaemia excesses which give cause for concern. For example, the Severnside Campaign Against Radiation has uncovered an excess of young leukaemia and lymphomas in the area opposite the nuclear river side plant[12] of Berkeley. This was later confirmed by the local health authority.[13] Six cases were found from 1979 to 1984 when less than one would be expected—another example of a super cluster near a nuclear installation. Since data on leukaemia deaths represent only about 30 per cent of all the information that can now be collected on leukaemia and lymphomas incidence the opportunity for finding further significant excesses exists and beckons.

CONCLUSIONS

Incidence data for leukaemia cases under 25 in Scotland and mortality data for

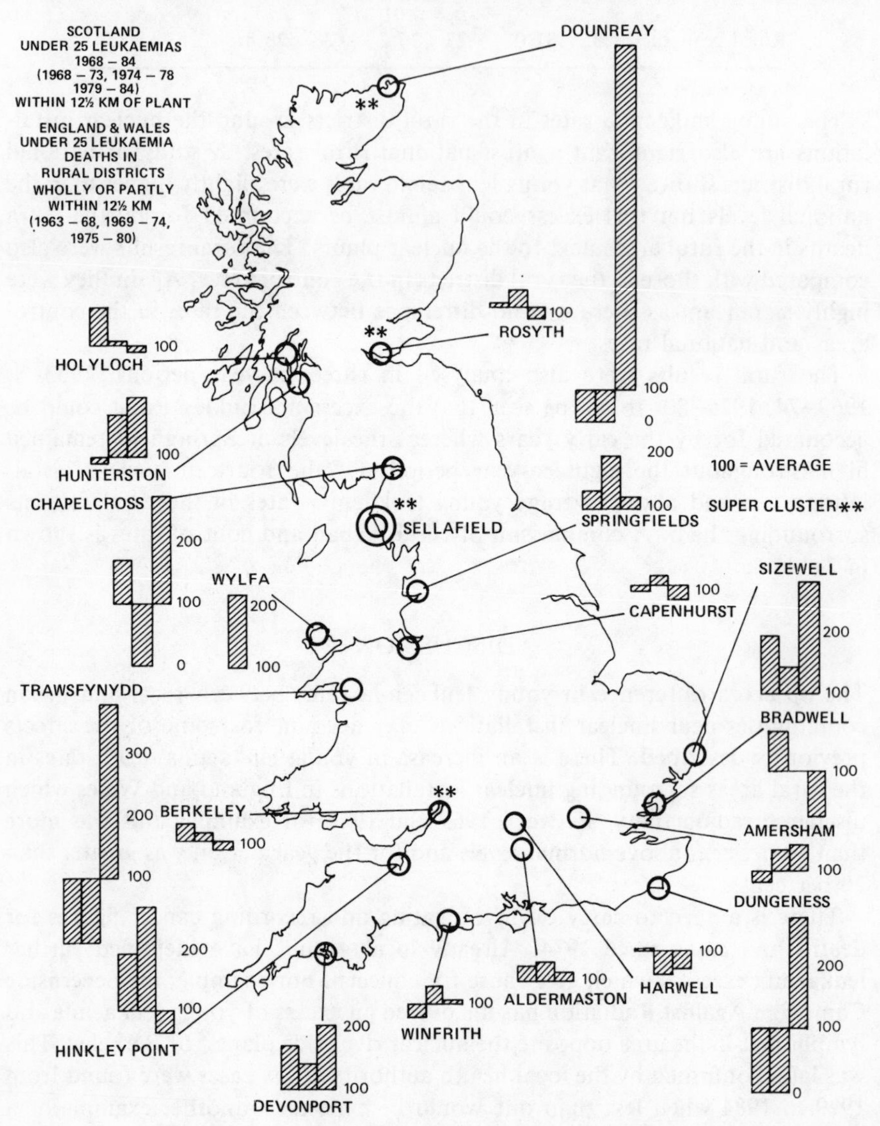

Figure 1 Leukaemia and nuclear power in Britain.

England and Wales is summarized graphically in Figure 1 for all major nuclear installations except Sellafield. The most important points are summarized below:

1. In 1983 there was the discovery of an unexplained excess of young leukaemia cases near the Sellafield nuclear reprocessing plant.
2. This excess lay outside the normal Poisson distribution and could be defined as a super cluster.
3. The Black report confirmed an unusual excess of leukaemias and lymphomas near Sellafield.
4. The hypothesis that average leukaemia rates existed round nuclear power plants was tested by the Scottish Health Service. Their findings demonstrated a further super cluster at Dounreay, and defined in terms of equivalent urban areas another super cluster at Rosyth.
5. In all other areas the clusters found in Scotland could be accounted for by random clumping but there were raised levels around nuclear installations.
6. A comprehensive survey of leukaemia death rates round fourteen nuclear power plants in England and Wales was carried out for the period 1963 to 1980.
7. A significantly raised mortality rate was found in rural areas compared with control areas and national rates of childhood leukaemia.

Finally it should be recognized that this information is based on publicly available documents. Scottish data can be obtained from the Dounreay Public Inquiry, c/o Department of Environment, 2 Marsham Street, London. England and Wales data from OPCS, St Catherine's House, Kingsway, London, where it has been collected and stored for the last 22 years.[14]

REFERENCES

1. J. A. Urquhart, M. Palmer, and J. Cutler, Cancer in Cumbria: the Windscale connection, *Lancet*, 28 Jan. 1984, 217–18.
2. *Investigation of the Possible Increased Incidence of Cancer in West Cumbria*, Report of the Independent Advisory Group, Chairman Sir Douglas Black, HMSO, 1984.
3. A. W. Craft and S. Openshaw, Apparent clusters of childhood lymphoid malignancy in Northern England, *Lancet*, 14 July, 1984, 96–7.
4. A. Pomiankowski, Cancer incidence at Sellafield, *Nature*, **V**, 311, 13 Sept, 1984, 100.
5. M. A. Heasman, I. W. Kemp, J. D. Urquhart and R. Black, Childhood leukaemias in Northern Scotland, *Lancet* 1 Feb. 1986, 266
6. Leukaemia and Young Persons in Scotland, Scottish Health Service, Evidence to the Dounreay Inquiry
7. R. A. Cartwright and J. G. Miller, Lymphoid and Haemopoietic malignancy case occurrence in U.K., 1984, *Lancet* 1 Dec. 1984.

8. G. J. Swansbury, Clustering of lymphoid and haemopoietic malignancy, *Lancet*, 16 Feb. 1985, 398.
9. R. A. Cartwright, Recent epidemiological studies of leukaemia in the United Kingdom, in *Epidemiology of Leukaemia and Lymphoma*, pp. 675–82. Leukaemia Research Special Issue, 1985.
10. J. A. Urquhart and J. Cutler, Incidence of childhood cancer in West Cumbria, *Lancet*, 19 Jan. 1985, 172.
11. A. W. Craft and S. Openshaw, Childhood cancer in West Cumbria. *Lancet*, 16 Feb. 1985, 403–4
12. Nuclear Power and Health in Gloucestershire, Severnside Campaign against Radiation, October 1986.
13. J. M. Stuart, *Childhood Cancer in Gloucestershire*. Report to Gloucester Health Authority, January 1987.
14. S. D. 25's for rural and urban districts in England and Wales 1963–1980.

Radiation and Health
Edited by R. Russell Jones and R. Southwood

19
Discussion Period 3

Dr Jacobs (*Haematologist, University Hospital Wales, Cardiff*) Dr Lewis, when you find people within a population who are unduly sensitive to ionizing radiation, are cells other than fibroblasts affected?

Dr Lewis We have been trying to evaluate a test for individual radio sensitivity which is based on the response of peripheral blood lymphocytes, looking at DNA reaggregation after disaggregation. It is a test which we are trying to validate by doing parallel studies on fibroblasts. We are anxious to evaluate this test because it can be performed in the course of a few days, in contrast with the fibroblast studies which take several weeks. It is important to develop a test whereby individuals can be examined quickly for their radiosensitivity, particularly in circumstances of industrial exposure.

Question Dr Lewis, you mentioned that in ataxia telangiectasia there is a predisposition to lymphomas and in xeroderma pigmentosum, there is a predilection for skin cancers. Is there any evidence of a generalized risk of cancers at other sites?

Dr Lewis The answer is no. In xeroderma pigmentosum many patients live to middle age and to my knowledge there is no definite evidence of increased visceral cancer. In ataxia telangiectasia, again there is only anecdotal evidence of excess cancers in other organs. But I am not convinced that any of these are significant.

Dr Russell Jones Professor Evans, in your studies on the Rosyth dockyard workers, you calculated that ten cigarettes a day produced about the same degree of chromosomal aberration as two to three rads per annum. However, in terms of cancer fatality ICRP predicts that five rads per year would be equivalent to about one or two cigarettes a day. So there is a tenfold discrepancy apparently, between your findings and the ICRP prediction.

Professor Evans That might not be surprising. I should add that my two to three rads is an approximation. It is not a calculation that is done to the nearest decimal point. I simply wanted to give some perspective on the effect of radiation as compared with smoking.

Dr Belbeoch Have you looked for synergistic effects?

Professor Evans We have conducted many experiments using x-rays in combination with other agents, particularly chemical agents. In most cases the effects are additive and not synergistic, the reason being that x-rays produce DNA damage in a very different way from chemical agents. X-rays actually damage the DNA directly. They produce free radicals which can cut the DNA. Most chemicals produce their effects by interfering with DNA replication.

Mr Webb Professor Evans, your comparison between different types of electricity generation workers. Was that study controlled for smoking?

Professor Evans Yes. It was controlled for smoking, for radiation and for infections.

Mr Webb Has any study been done on the effects of either alpha or beta exposures from internal isotope examinations, as opposed to external x-rays?

Professor Evans *In vitro* there has been a lot of work; by David Lloyd for example at the NRPB. There have been some studies *in vivo*, mainly on nuclear workers in the States. The problem of course is dosimetry.

Mr Webb So we do not have the same kind of dose–response relationship that you are able to show with your studies?

Professor Evans Not *in vivo*, we do *in vitro*. The response is linear. Alpha particles give you a straight line and so do betas.

Professor Scorer What worries me about your analysis is that it depends on cells from the peripheral circulation. How relevant is this to cancer risk?

Professor Evans We use blood lymphocytes because they are so readily accessible. It is painless for the person donating the cells and it is very easy to obtain lots of chromosomes. Getting a bone marrow is difficult and painful.

Dr Russell Jones Could I make a general point Professor Evans, about the comparison you drew between the number of chromosomal aberrations seen in radiation workers as opposed to smoking? Now this is a comparison that is very often made; usually by doctors who are involved in public health issues and who wish to reassure people who are, in their eyes, excessively concerned about the risk of ionizing radiation. However, there are major differences between the public perception of voluntary and involuntary risks. In the United States the Electric Power Research Institute researched this area and found that the risk that the public is prepared to accept, when it is a risk that is imposed upon them, carries a health risk that is one thousandth of the risk that they are prepared to run if they are actually doing it to themselves. Now of course you may find that finding rather peculiar or extraordinary. But it is a reflection of the fact that the public do perceive these risks very differently, and that it does not carry that much weight with people when you tell them that smoking will do them far more harm.

Professor Evans That was not the thrust of my message. My message is that they all cause you harm. I do not think that radiation causes you more than smoking or vice versa. However, I think that cigarette smoking is a useful comparison because there is good information of a quantitative sort in relation to cigarette smoking and inherited mutations. Personally I do not want any of these carcinogens around. I want to live in an environment that produces least damage to my chromosomes.

Question I was interested in the linear relationship between chromosomal aberration and the exposure of nuclear dockyard workers.

Professor Evans In the nuclear dock-workers there was a dose response seen after ten years' exposure. Obviously cells have turned over in those ten years. So this is a different phenomenon from *in vitro* responses where there is a one-off radiation exposure and changes are observed in cells which have undergone only one cell division. The *in vivo* effect is linear.

Mr Stewart Boyle (*Friends of the Earth*) I would like to ask Dr Darby a question in relation to the relative contributions of fallout and discharge. Our knowledge of nuclear discharges is by no means complete, particularly in the 1950s and 1960s. During cross-examination, results of soil sampling were presented to the Dounreay Enquiry. The number of samples was limited, so we have to be careful how much weight we attach to them. Even so, the level of caesium-137 in the soil was greater than would be expected from fallout, by a factor of four or five.

Dr Darby I was not privileged to be at the Dounreay Enquiry, but I am sure there are lots of people in the audience who were.

Dr Russell Jones I actually presented that particular data to the Dounreay Enquiry. Eventually I obtained from the NRPB, the level of caesium-137 that they were expecting in soil as a result of the Dounreay discharges. At a distance of 6 kilometres from the plant, the NRPB predicted an excess over fallout of 0.25 of a becquerel per kilogram. The excess that we noted in the soil sample was 67 becquerels per kilogram, so there is a 250 discrepancy between the observed and the expected which reinforces the point that it is very unwise to rely entirely on mathematical modelling unless you have samples which reinforce your predictions.

Dr Darby Oh, I entirely agree. I think that as much sampling as possible should be carried out.

Professor Jacobi I have a question for Dr Stewart. I assume that the exposure rates which you have taken from the NRPB refer to the dose in outdoor air and we know that indoors the terrestrial dose rates are quite different.

Dr Stewart You are quite right, we took the outdoor dose, but if you were here for the lecture by Dr Fry, you will have noticed that the indoor dose taken nationwide follows very closely with external. There is no contrast there.

Professor Jacobi But there are still differences between outdoor and indoor doses depending upon the type of building material used?

Dr Stewart Well if this was a major factor we should have got a negative finding. In fact, we have a positive finding, therefore it cannot be disturbing it too much.

Dr M.Lesna (*Pathologist, Royal Victoria Hospital, Bournemouth*) Dr Stewart I would like to hear more about those other factors which relate to childhood cancer. I am particularly interested to learn whether mothers have developed tumours subsequently, or was there anything relevant in the occupational histories of the parents?

Dr Stewart There is a large literature on this subject. The smoking was only analysed for a limited period and nothing significant was detected. As for familial cancers, they are very rare and would not disturb the overall findings. George Kneale may wish to comment.

Mr George Kneale (*Cancer Epidemiology Research Unit, Birmingham*) We have got data on smoking, but only for about five years after death, because the question was only asked in the questionnaire at that time. There appears to be some kind of bias in the recording of smoking because more mothers of cancer children stopped smoking during pregnancy than control mothers. On the other hand, the number who smoked out of pregnancy is higher so the overall picture is a bit confused.

Dr Russell Jones The situation at the moment is that four of the most significant leukaemia clusters so far identified in the UK are around Sellafield, Dounreay, Rosyth and Berkeley. Two of these are reprocessing plants which discharge more radioactivity into the environment than the other nuclear facilities in the UK. So if one was going to look for a biological effect, that is precisely where you would expect to find it. Now two years ago Sir Douglas Black, when you produced your report, you only had the benefit of knowing about one of these clusters. Do you have now, any reason to advance the position that you held two years ago?

Sir Douglas Black Well obviously one keeps an open mind on this thing as new evidence comes in. And I think that it has to be said that it is disquieting that we are now learning things about Sellafield which did not know at the time we completed our report. Being all cynics we naturally mentioned that in the report as a possibility, and it is slightly disquieting that its now come to light. But if I can, since I was asked the question, the latest thing that happened there was that some radioiodine had escaped some fifty years ago and the knowledge of that leaked out through the Fifty Year Rule, and I was rung up by someone from radio and I said what isotope was it and he said radioiodine. I then asked him how much there was and he said he did not know and then I said, well, it is perhaps a little unwise to ask me to comment on a radioisotope which has not much to do, as far as is known, with leukaemia, and whose amount is unknown. And he became very sad and regretful and he said you have killed my interest in that story.

Mr Clark Bullard (*USA*) Mr Urquhart, I am concerned by your statement that the control areas are upwind, since the wind will not blow in the same direction all the time. Secondly, your results imply that either the dose estimates are wrong, or the models are wrong. In your opinion which do you think it is?

Mr Urquhart I notice, sir, that you come from the USA where the dimension of the problem is rather different. Your nuclear reprocessing plant, Hanford, for example, covers 749 square miles, which is the equivalent of the whole of West Cumbria. But we do have a problem. We are a small island and the question of control areas is crucial. In this study they were selected on the basis of distance away from nuclear facilities, that is 20 miles.

As for your second point, are the models wrong or are the dose estimates wrong, my response is that statistics cannot really answer that question. What they can do is to challenge the received wisdom and to suggest areas where further research should be carried out.

Dr M. Snee (*Medical Research Council, Southampton General Hospital*) Could I ask Mr Urquhart why the boundary of 12.5 kilometres is chosen, because that seems rather arbitrary.

Mr John Urquhart Well, first of all, if you look at the article in *The Lancet* which reported the Dounreay cluster, they chose 12.5 kilometre circles and 25 kilometre circles. Even within 25 kilometres there was a significant increase. Second, if you look at the postcode sector analysis that was provided to the Dounreay enquiry, then in the 1974–78 period, the worst postcode for leukaemia contained the Dounreay nuclear reprocessing plant. Now, you are quite right to say that you can go on drawing circles wherever you like. There are two ways of getting round that. One is to do an analysis which is boundary free. The other way is to take the 12.5 kilometres radius and test it at other localities.

Dr Taylor Could I clarify what exactly is a cluster? At the moment a cluster seems to be due to geographical clustering over a period of time; 5, 10, 20 or 30 years. In your report Sir Douglas, the geographical cluster also had a cluster within a cluster, in that if you looked at the dates of birth of the children who had died, then they clustered around 1957/1958 and indeed of all the children who died, about half were either *in utero* or in the first year of life at the time of the 1957 fire. Now that fire not only released radioactive iodine which we knew about at the time, but it also released radioactive polonium and the behaviour of that in the human system is not well known. I wonder whether there may be similar clusters within a cluster in relation to the Dounreay data. Have you looked at the dates of birth of the children who died in Dounreay and does that correlate with any events recorded, or even unrecorded, at Dounreay?

Dr Darby Well, I do not think I have used the word cluster until now. Point number 2, I think it is rather hard to speculate as to whether the events correlate with other unrecorded events. Point number 3, so far as I know, there is no obvious correlation between recorded events and the dates of birth.

Dr Stewart If we are right about all children's cancers having been 'triggered' *in utero*, then you must do it by date of birth. Anything afterwards may be too late. What about that?

Sir Douglas Black I am not going to say what about that! But we used the word cluster only to dismiss it, because we felt that it needed definition both in space and in time and we did not think that our data in fact showed that. Graphically the leukaemias are fairly evenly spread throughout the whole period. I take the point about time of birth, but even that did not stand up for all of the cases.

Dr Stewart But surely if we are right in what we said, then a cluster is contained within a defined population of children who were born at a given time, and by that date they have already acquired their induced cancers. You follow them over the next sixteen years, if they exhibit an excess number of deaths, that is a cluster. It is a little epidemic among that cohort. It allows also for the fact that not every case is going to have exactly the same latency. It does not advance matters by saying there are not clusters; there are clusters. You can call them mini-epidemics if you like, but it says that there are some birth cohorts that suffered unduly and some have got off scot-free.

Mr Urquhart Like Dr Darby I dislike the term cluster. In fact, in the Yorkshire TV programme, we used the word excess, and since then the word has been corrupted in the way the word actress has—all actresses are now stars, so you have to use the word superstar. It is the same principle. You have the word supercluster to try and distinguish excesses which are outside of the normal Poisson distribution.

I think Dr Stewart's point is quite right. There are excesses that are unaccounted for. That is the way of looking at it and logically it is important to look at one cause at a time. We ought to get the nuclear thing out of the way and then we can start looking for other causes.

Part 5
Chernobyl

Radiation and Health
Edited by R. Russell Jones and R. Southwood

Published 1987 by John Wiley & Sons Ltd.

20

Dose Distributions in Western Europe Following Chernobyl

ROGER H. CLARKE
Secretary, National Radiological Protection Board, Chilton, Didcot, Oxford, UK

INTRODUCTION

The reactor accident at Chernobyl Unit 4 which began on 26 April 1986 led to large quantities of radioactive material being released to the atmosphere over the following ten days. During this time there was a complex set of meteorological conditions over Europe which dispersed and deposited these materials from Scandinavia to Turkey. Levels of radioactive material in the environment were monitored by national authorities and in some countries, where the levels were sufficiently high, measures were taken to reduce the radiation exposure of the highest exposed groups in the population.

In this chapter I describe the different circumstances that, in a reactor accident, lead to varying compositions of radionuclides being released, depending on the operational state of the reactor and the exact sequence of events. The range of radionuclides released is so variable that it is not possible to know in advance either what will be released or what pathways of exposure will be important. I illustrate this with a comparison of the magnitudes of release and radiological consequences of the 1957 Windscale accident, that at Three Mile Island in 1979 and Chernobyl in 1986.

I will then trace the atmospheric dispersion of material from the Chernobyl accident and explain the differences in deposition that occurred across Scandinavia and Europe. This leads to estimates of ranges of doses in each country and of contributions from each exposure pathway. Finally I will mention the radiological basis for taking action to protect members of the public and the difficulties encountered in practice when applying the criteria.

THE CONSEQUENCES OF A REACTOR ACCIDENT

When uranium undergoes fission, fission products are formed which cover a range of elements from zinc through strontium and iodine to barium and cerium. For each of these elements there are formed radioisotopes with varying half-lives, from fractions of a second to tens or hundreds of years and occasionally more. The actual quantities of each of these radioisotopes which will be present in a nuclear reactor at any point in time will depend on the yield of that radioisotope per fission event, the power level of the reactor and the length of time for which the fuel has been irradiated.

Long-lived fission products such as Cs-137 with a half-life of 30 years build up linearly with fuel irradiation, while short-lived ones such as the 8-day half-life I-131 achieve an equilibrium depending on power level. Thus the nuclides available for release in an accident will depend upon the power rating of the fuel, its length of burn-up, and whether the reactor had been operating at full power for a time or had recently been started up after shutdown.

Although there will be a range of radionuclides available for release which depends on operational history, the quantity released will depend upon the temperatures to which the fuel is raised in an accident sequence and the chemical species of the fission products—their volatility. If the fuel cans fail then the first fission products to be released will be krypton and xenon. The volatile fission products (iodine, tellurium) are likely to be released at a rate about one-tenth that of the noble gases, the mid-volatiles (ruthenium, caesium) at a rate about one-hundredth of that of the noble gases, and the non-volatiles (rare earths) at about one-thousandth of the rate of noble gases.

An explosive disruptive event may lead to fuel vaporization in which case the full spectrum of radionuclides in fuel will be released. The actual amounts of radionuclides that will emerge into the atmosphere will depend on any plate-out of radionuclides on surfaces over which they pass in their pathway from core to atmosphere.

I can illustrate these effects by comparing the 1957 Windscale accident in the UK with the Three Mile Island accident in the USA in 1979 and that of Chernobyl this year. In Table 1 I have compared the quantity released of the principal nuclide of radiological concern in each of the accidents with the doses to the most exposed individuals and an estimate of the total radiological impact of the accident. The Windscale accident released principally the short-lived I-131; there was little Cs-137 and few other long-lived radionuclides in the fuel because it was a very low burn-up fuel cycle.

In the TMI accident, although high burn-up fuel was involved the containment did not fail, the fission products were retained in the vessel and while water escaped from the core through the faulty valve only the gas Xe-133 escaped to atmosphere. The quantity of activity released was more than 400 times greater than from the Windscale accident but the effective dose to the most exposed individuals was ten times lower. This is because of the difference in pathway of exposure: a cloud-γ exposure compared with I-131 in cows' milk. Expressing results in effective dose term allows comparisons to be made when different organs are irradiated either individually or in combination. For

Table 1 Comparison of releases and impacts of reactor accidents

	Release (Bq)	Principal Nuclide	Maximum off-site dose (Sv)	Collective effective dose (man Sv)
Windscale 1957	$7\ 10^{14}$	I-131	0.16^a 0.009^b	2000
TMI 1979	$3\ 10^{17}$	Xe-133	0.001^b	33
Chernobyl	$2\ 10^{18}$	I-131 Cs-137 mixed fission products	2.5^a 0.25^b	$2\ 10^5$–$2\ 10^6$ (USSR) 3000 (UK)

[a] Thyroid
[b] Effective

Chernobyl, the release was ten times higher than TMI but the doses were several hundred times higher. Most of the Chernobyl dose arises from I-131 and Cs-137 as we will see later in the chapter.

The total impact, as expressed by collective dose, was greater in the UK from the Chernobyl accident than from the Windscale accident (3000 cf 2000 man Sv) while the TMI total impact in the USA was 100 times less. The impact of the Chernobyl accident in the USSR has been estimated by the Soviets at 2 million man Sv, but the IAEA experts have suggested it may be ten times lower[1], the discrepancy being due to differing views on the persistence of Cs-137 in the environment.

The conclusion that I want to draw at this point is that different accident sequences may lead to differing radionuclides being released and various pathways may be important. In planning to deal with an accident it is necessary to prepare for a wide range of radionuclides and a wide range of pathways of exposure; external radiation from the plume; inhalation of activity in the plume; external dose from ground deposits of activity; and consumption of contaminated foodstuffs.

RELEASES FROM CHERNOBYL

The accident at Unit 4 of the Chernobyl nuclear power station occurred on the 26 April 1986 during a test of the ability of one of the turbines to supply the reactor's power requirements should a power failure ever occur[1]. However, a series of human errors, whereby safety systems were deliberately switched off and operating rules were ignored, brought the reactor into an unstable condition in which rapid boiling of the coolant occurred and led, for this reactor design, to an increase in power to a hundred times normal full power in a

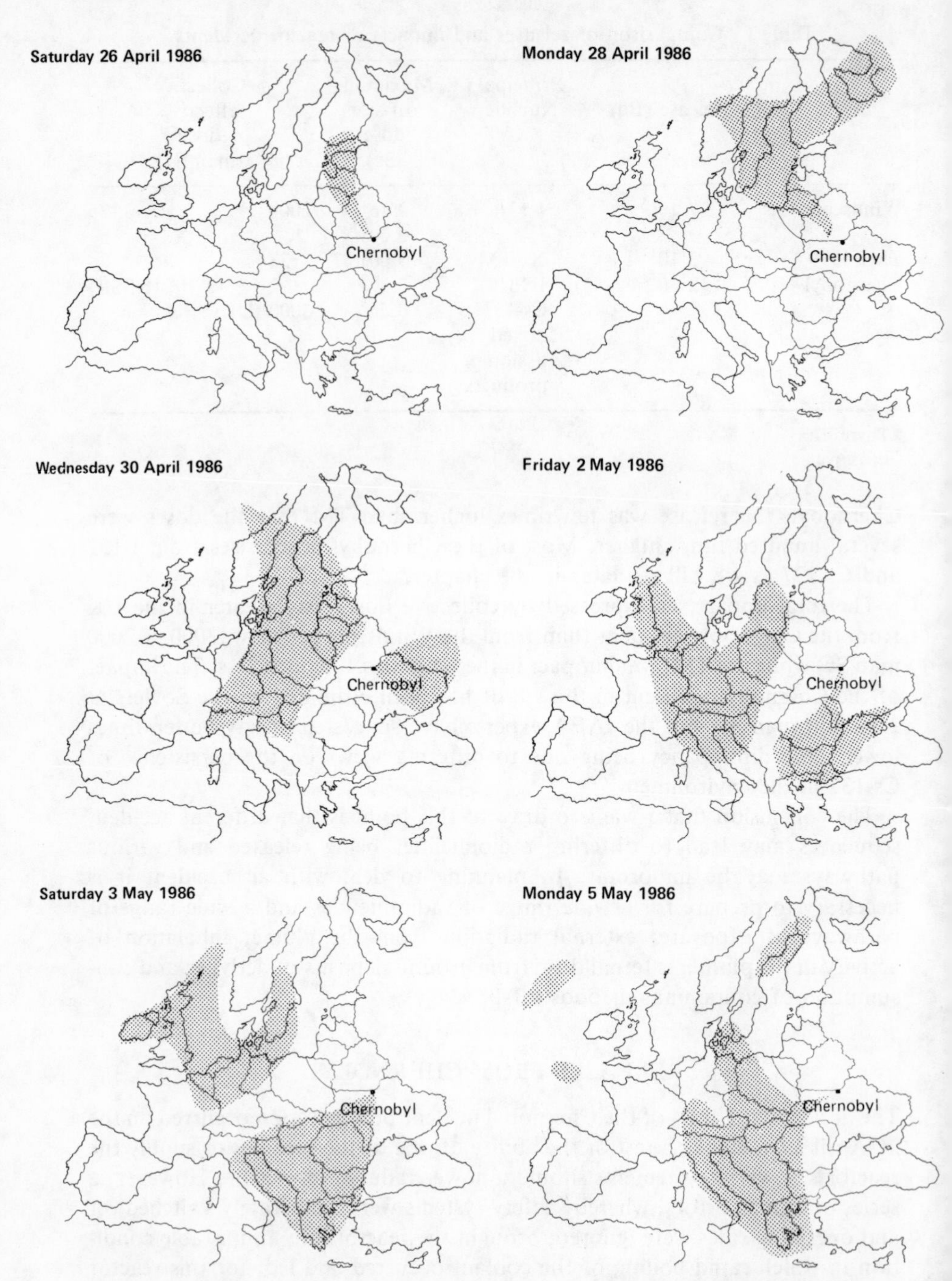

Figure 1 Areas covered by the main body of the cloud on various days during the release.

matter of a few seconds. This increase in power occurred at 0123 h local time on the 26 April (2123 h GMT on the 25 April), caused a steam explosion which shifted the 1000 tonne top cover off the reactor. After 2–3 seconds, a second explosion occurred and hot pieces of the reactor were ejected from the destroyed reactor building. A mixture of radioactive gases and particulate matter was carried to heights exceeding 1200 m.[1]

The release of radioactive material to the atmosphere continued over a period of about ten days with two peaks in release rate, on the first day (26 April) and on the tenth day (5 May) during which time mitigating actions by the Soviets brought the reactor into a state where further major releases were unlikely.[1]

Figure 1 shows estimates of the areas of land covered by the main body of the plume at various times during and after the release. The figures are based on trajectories calculated by the UK meteorological office and included in a report by NRPB for the Commission of the European Communities.[2] Figure 1(a) shows that on Saturday, 26 April, the day of the accident, only areas to the north of Chernobyl were affected by the plume, but that by Monday (Figure 1(b)) areas of Scandinavia and north-east Poland were affected. Radiation readings on the eastern coast of Sweden were fourteen times background, indicating that a plume of radioactivity had reached Scandinavia, and alerting the international community that a large accident had occurred. By the Wednesday (Figure 1(c)) the wind direction at the site had changed leading to a plume of material travelling to the south and east. A high pressure system developed causing the contaminated air mass to split, spreading activity over other regions of Europe.

By Friday the 2 May (Figure 1(d)) the initial contamination reached the UK, while material released at that time from Chernobyl was moving southwards over Greece. By Saturday 3 May (Figure 1(e)) two main areas of contaminated air existed, one over north-western Europe and one over south-eastern Europe. On Monday 5 May, when a second major peak in the release rate from the damaged reactor occurred, the main contaminated air mass lay over southern Germany, Italy, Greece and eastern Europe, with the remains of the initial contaminated air mass dispersing over the Atlantic. By Tuesday, 6 May the rate of release from the damaged reactor had fallen to relatively low levels.

PATHWAYS OF EXPOSURE

A wide range of radionuclides was measured in the air samples that were taken in all countries in Western Europe. The range of fission products includes [3,4,5] Zr-95, Nb-95, Mo-99, Ru-103, Ru-106, I-131, I-132, Te-132, Cs-134, Cs-137, Ba-140, La-140, Ce-141 and Ce-144. Measurements of actinides in Western Europe identified, Pu-238, Pu-239, Pu-240, Am-241 and Cm-242 in air, but at very low levels and having little radiological impact. The most important nuclides as far as dose to people was concerned were I-131, Cs-134 and Cs-137. It can be seen from Table 2 that the γ-dose from the passing cloud and inhalation of material from that cloud were probably the least important pathways of exposure for the population. The most important routes of exposure were those of ingestion and external radiation from deposited material. The figures are approximate since both habits and relative activity levels varied over Europe but illustrate the relative significance of pathways. It is possible that individuals existed whose dose contributions were outside this range because of the different levels of activity and dietary intake.

The group in the population liable to receive highest doses was identified as young children drinking fresh milk from animals grazing contaminated pastures. Other foods, such as leafy green vegetables and fruit, were identified as possible sources of exposure. Depending on the level of contamination found and the radiological criteria being applied, counter-measures were instigated by national authorities to limit radiation exposure. These counter-measures included, for example, requiring grazing animals to move from pasture to stored feed, instructing people to wash fruit and vegetables before eating them, advising people against drinking undiluted rainwater, and, in countries where the levels were particularly high, the withdrawal of fresh milk and vegetables from public consumption. These counter-measures remained in place for various lengths of time, ranging from a few days to a month in some places, again depending on the radiological criteria being adopted and on the environmental levels found.

As it became apparent that the level of the short-lived iodine-131 was not the only nuclide of concern, account had to be taken of the contribution from the longer-lived isotopes of caesium. In the early phase this meant taking account of both iodine and caesium isotopes in milk while later on, additional counter-measures were introduced to limit the long-term intake of caesium radioisotopes. Concern about fresh milk and fresh vegetables thus became

Table 2 Contribution by pathway to individual effective dose in the first year

Main pathways

	Percent contribution range[a]	
Pathway	Average adult	Critical group
Inhalation	<15	$\leqslant$5
Ingestion	55–90	70–98
Deposited gamma	10–35	1–25
Cloud gamma	< 0.3	< 0.3

[a] These percentages are approximate, and intended to illustrate the relative significance of pathways only.

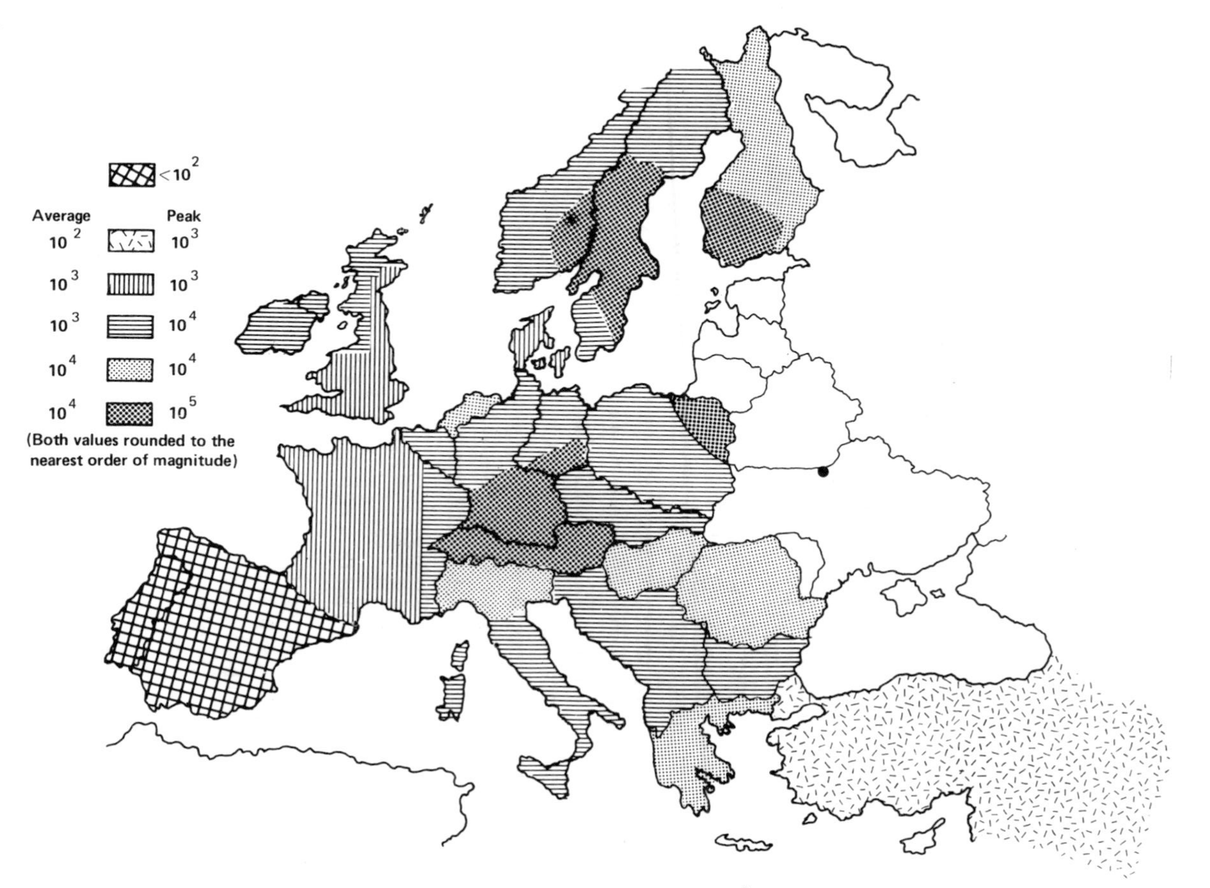

Figure 2 Iodine-131 deposition, Bq m^{-2}

reduced but other foods, such as meat and processed milk products, became of concern. Although a range of radionuclides have been detected in various environmental materials in all countries since the Chernobyl release, in all cases the most important radionuclides were iodine-131, caesium-134 and caesium-137.

LEVELS OF DEPOSITION

Figure 2 shows the broad variation in the total deposit of iodine-131 throughout Europe and Scandinavia. The highest levels of deposition occurred where there was heavy rainfall during the passage of the radioactive plume. The highest levels of iodine-131 contamination were seen in Scandinavia, southern Germany, Austria, Switzerland, Northern Italy and Greece. Similar patterns of deposition of caesium-137 were found, as illustrated in Figure 3. However, there are some differences, due to the differing behaviour of the two radionuclides. Figures 2 and 3 are based on the available environmental data from each country and inevitably I have used some judgement in producing these broad figures. I think they are a fair representation of the general situation in each country or region but areas within each region could have higher or lower depositions depending on local factors.

In those areas where it was dry while the plume was passing, the deposition of caesium isotopes was less, by up to a factor of 10, than that of I-131 although the exact amount depended on the relative concentrations of I-131 and Cs isotopes in the plume. Where it rained through the contaminated air mass, I-131 deposition was greater than in dry areas and the deposition of Cs isotopes was similar to that of iodine.

RESULTS OF THE DOSE ASSESSMENT

The average effective doses to adult individuals in each European Community country from exposure in the first year following the Chernobyl accident are shown in Table 3. These doses are summed over all radionuclides and exposure pathways and were calculated taking account of any counter-measures said to have been imposed. Where different regions within a country were considered, the average dose was obtained by taking a weighted mean of the results for each region, based on population in the regions. The average doses range from 0.2 μSv in Portugal to between 200–400 μSv in the FRG, Italy and Greece. Sweden, Finland, Switzerland and Austria are likely to be similar to the FRG. Consumption of food and external irradiation from deposited material are the major contributors to doses in the first year. Differences between countries in the relative contributions of these two pathways are partly due to the counter-measures adopted but mainly to differences in deposition and the food pathways back to man. The fifty-year integrated dose shown in Figure 4 in

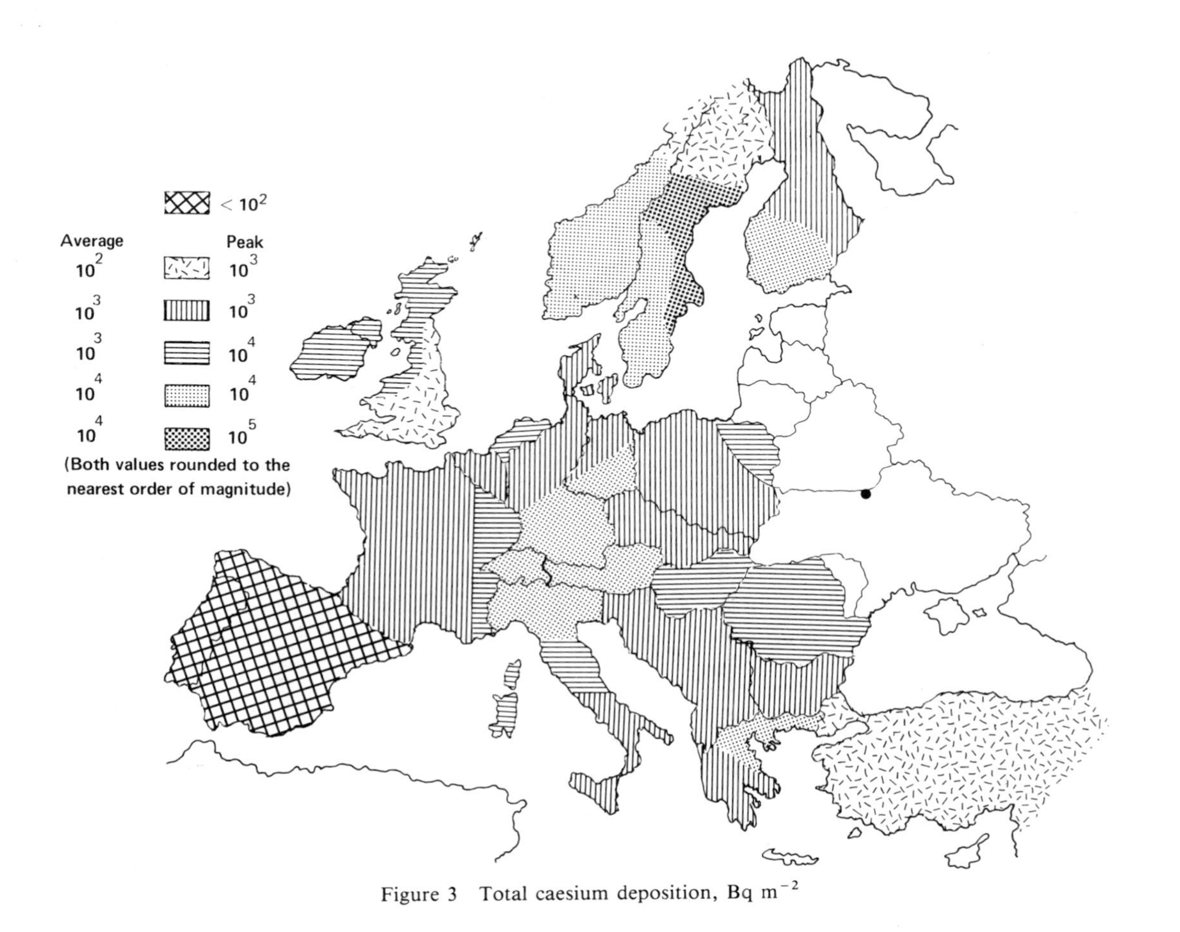

Figure 3 Total caesium deposition, Bq m^{-2}

Table 3 Average adult effective doses in the first year and integrated to 50 years

Country	Effective dose equivalent (μSv)	
	First year	50 years
Belgium	51	90
Denmark	63	100
Eire	100	170
France	38	74
FRG	190	410
Greece	350	580
Italy	210	370
Luxembourg	61	100
Netherlands	68	110
Portugal	0.2	0.3
Spain	1	1.2
UK	33	48

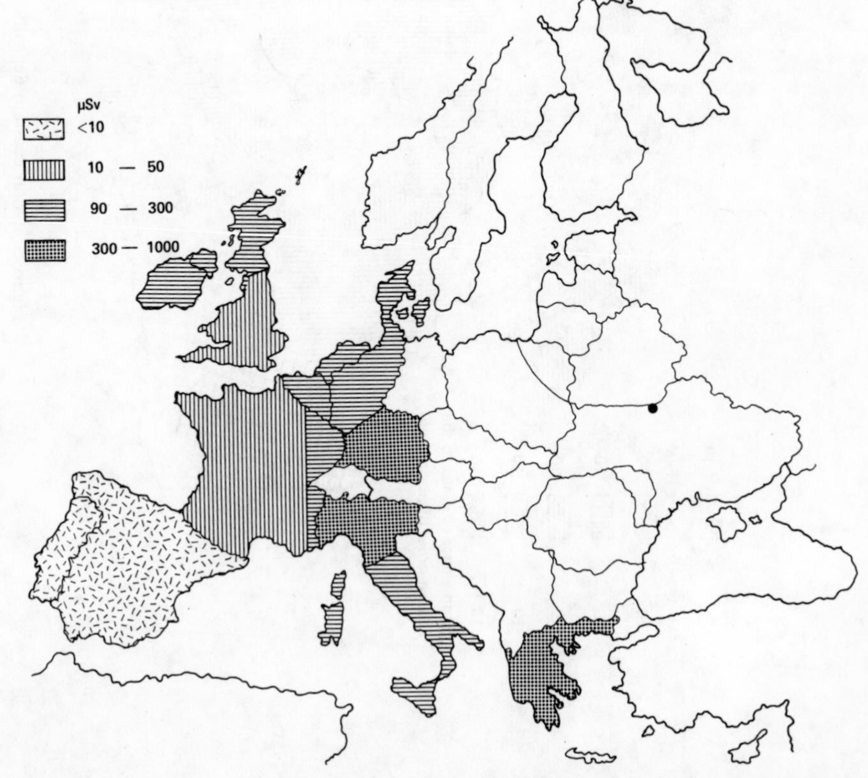

Figure 4 Average adult effective dose to 50 years within the EC with counter-measures.

broad dose ranges is between a factor of 1.5 and 2 higher than the first year dose depending on the relative ratios of I-131 and Cs isotopes deposited on the ground (Table 3). The figures for dose are lower than those in early estimates[3] because of the more comprehensive environmental monitoring data now available.

Doses to the thyroids of individuals in the first year after the accident will be higher than the effective dose. However, the objective in defining effective dose is to be able to express irradiation in terms of the same level of fatal risk, regardless of whether the whole body is irradiated or only partial irradiation occurs.

The average effective doses may be placed in perspective by comparing them with the effective dose received each year from natural background radiation. In Europe, average individual effective doses from natural background are about 2000 μSv per year. Thus doses to individuals over the 50 years following the Chernobyl accident will be much less than those received in one year from natural background radiation. Alternatively it may be noted that a return flight by jet to Spain from the UK incurs an effective dose of about 20 μSv. This is similar to the dose from Chernobyl over much of England and France.

Effective doses to critical group individuals in each country, in the first year after the accident[2] range from five to about 20 times the average. The critical group of individuals may be the 1-year-old infant, the 10-year-old child or adult depending on the country concerned, and generally there is little difference in the doses predicted for the three age groups. The most important exposure pathway in all cases is ingestion of contaminated food. Table 4 shows the contribution to dose made by each foodstuff. For infants milk is generally

Table 4 Contribution of individual foods to dose

Food	Per cent contribution range[a,b]	
	Average adult	Critical 1-year-old infant[c]
Milk	20–60	15–90
Milk Products	< 10	< 5–20
Grain	5–60	< 1–10
Lamb	10–35	< 1–30
Green vegetables and fruit	1–30	10–30
Beef	5–25	$\leqslant 5$

[a] These percentages are approximate, and intended to illustrate the relative pathways only.
[b] The most important foods vary between countries. Those given here are the ones which are generally important throughout Europe, but all the ranges may not be directly applicable in some member states.
[c] Because eating habits vary considerably between age groups, the contributions of foods for 1-year-old infants with extreme habits is given.

the most important food but for the other age groups milk, lamb, green vegetables and fruit all make significant contributions to the critical group dose.

COLLECTIVE DOSE ASSESSMENT

Estimates of the collective effective doses to the populations of EC countries in the first year after the accident, and the collective effective doses integrated over all time, i.e. the collective effective dose commitments, have been produced[2].

The collective effective dose commitment to the whole population of the EC is estimated to be 80 000 man Sv. The highest collective effective doses are those to the populations of the FRG (30 000 man Sv) and Italy (26 000 man Sv), because contamination levels in these countries were relatively high and the populations of the FRG and Italy are relatively large. The main pathways contributing to the collective effective dose are consumption of contaminated food and external irradiation from deposited material. In the first year food consumption is the dominant pathway, but in later years external irradiation from deposited material increases in importance. For both pathways, caesium-137 is the most important radionuclide.

The total collective thyroid dose to the EC population is estimated to be 180 000 man Sv. The dominant radionuclide in this instance is iodine-131 and since this has a relatively short radioactive half-life (8 days), a large fraction of the dose would have been received within a few weeks of the accident. For thyroid doses, ingestion is the most important pathway.

The estimated collective effective dose in the EC from the Chernobyl accident, integrated over all time, is about 16 per cent of the *annual* collective effective dose from natural background radiation, some 500 000 man Sv.

Since in radiation protection it is assumed that dose is linearly related to risk without threshold it is possible to calculate the number of theoretical additional thyroid cancers occurring within EC countries due to Chernobyl to be some two thousand, of which about 5 per cent are expected to result in fatality. The number of thyroid cancers expected in EC countries over the next thirty years, even if the Chernobyl release had not happened, is of the order of one or two hundred thousand. The number of additional theoretical fatalities from cancers of all types due to Chernobyl can be predicted to be in the region of a thousand. These extra cancers are predicted to occur spread out in time over a few decades following the accident. Over the next thirty years about twenty million or so people in the EC countries are expected to die from cancer of one type or another.

The conclusion that I want to draw here is that if the present risk models are correct, we will never see any effects of Chernobyl from any epidemiological studies that might be established in Western Europe.

LESSONS LEARNED

The first lesson was that the demand for monitoring was overwhelming. This included monitoring of the environment, monitoring of people returning from the USSR and Eastern Europe, monitoring of aircraft flying across Europe, monitoring of foodstuffs and other cargoes at ports. Had the accident occurred in Western Europe the level of demand for these measurements might well have exceeded available resources.

The second lesson was that better communication with the public is required. There is a need to explain why people are expected to continue living in any enhanced artificial radioactive environment: although the doses may be small, the activity is measurable. The radiological protection assumption of linear non-threshold dose–response means there is no level of zero risk and there is a need to argue the acceptability of the assumed risk.

Thirdly there is a need to consider whether there can ever be unified 'action levels'. There was a general agreement in Europe on the level of dose which would justify action on radiological grounds. This level was that set by ICRP[6] and WHO.[7] Decision making had to move from an acute irradiation situation of ^{131}I in milk or fresh vegetables, through an acute situation that included Cs-134 and Cs-137 as well as I-131, to a chronic exposure situation as it was realized that ^{137}Cs and ^{134}Cs in processed foodstuffs, e.g. powdered milk, tinned or frozen vegetables or meat (particularly lamb), could be a source of dietary intake over a year. This resulted in lower 'action levels' in terms of Bq kg^{-1} for these processed products to give the same level of dose as from fresh items giving short-term intakes of radionuclides.

Finally, radiological protection concepts proved difficult to explain to the public, politicians and administrators. The radiological protection community must decide which way to go in the future—either we need to work hard to educate people to understand our philosophy, or we accept that it is politically expedient, but expensive, to choose single, simple arbitrary numbers for activity concentrations in food and water.

REFERENCES

1. International Atomic Energy Agency. *Summary Report on the Post Accident Review Meeting of the Chernobyl Accident.* Safety Series 75-INSAG-1, IAEA Vienna, 1986.
2. M. Morrey, J. Brown, J. A. Williams, M. J. Crick, J. R. Simmonds and M. D. Hill, *A Preliminary Assessment of the Radiological Impact of the Chernobyl Reactor Accident on the Population of the European Community.* CEC Report, Jan. 1987.
3. F. A. Fry, R. H. Clarke and M. O'Riordan, Early estimates of UK radiation doses from the Chernobyl reactor. *Nature*, **321**, 15 May 1986.

4. C. Hohenemser, M. Deicher, *et al*. Agricultural impact of Chernobyl: a warning. *Nature*, **321**, 26 June 1986.
5. A. J. Thomas and J. M. Martin, First Assessment of Chernobyl radioactive plume over Paris. *Nature*, **321**, 26 June 1986.
6. International Commission on Radiological Protection. *Protection of the Public in the Event of Major Radiation Accidents: Principles for Planning*. ICRP Publication 40, Pergamon Press, 1984.
7. *Nuclear Power: Accidental Releases—Principles of Public Health Action*. World Health Organization, Regional Office for Europe, Copenhagen, Publication 16, 1984.

Radiation and Health
Edited by R. Russell Jones and R. Southwood

21
The Effects of Chernobyl

B. E. LAMBERT
Department of Radiation Biology, Medical College of St Bartholomew's Hospital, London, UK.

The effects of the nuclear catastrophe at Chernobyl have been widespread in Europe and are of relevance to this conference because much of the radiation was delivered outside of Russia. The effects will not just be on health; economic and power policies have been affected and there could therefore be an eddy to disturb the social fabric. Consideration of these multitudinous effects is worthwhile because it illustrates deficiencies in the system of planning for such accidents and underlines the apprehension in the public's attitude towards nuclear power. If nothing else the whole sorry mess demonstrates to the public that we were unprepared for such an emergency, even 2000 miles away, and that there was no general consensus among informed scientists in Europe about what to do and when. Fuelling the public's unease is the history of secrecy in the nuclear industry which has only recently been lifted. However, the French authorities who have more to lose than any other European country if the concept of energy generation by nuclear power is threatened, denied the existence of Chernobyl fallout for a time. Such actions inspire no confidence.

However, to salvage some good from this accident we should consider what actually occurred in the UK and learn from the mistakes made. It is also of value to discuss the state of knowledge and uncertainty in the science which enables predictions to be made of long-term health effects. This may then focus attention on the system of dose limitation and allow discussion of the adequacy of radiological protection afforded to both workers and members of the public.

HEALTH EFFECTS

In simplistic terms about 3–10 per cent of the core of the Chernobyl 4 reactor was burned off on 26 April and four days later the resulting cloud of debris reached the UK. For about the next ten days the people of the UK were exposed to external radiation and for a longer period to the intake of contaminated food and water. Monitoring of the environment, food etc. was carried out and, on the basis of somewhat tenuous models, theoretical doses were calculated. The first of these estimates for the UK, by the NRPB, were published in May.[1] On the assumption of no threshold even these relatively small doses can be expected to produce some long-term effects in terms of increased incidence of cancer and very soon estimates of these effects were made. Estimates varied widely depending almost entirely on the particular risk function chosen. Thus increased numbers of cancers predicted in the UK varied from about fifty by the NRPB to more than 500. The most extreme estimates have been made by von Hippel and Cochran[2] who predicted up to about 120 000 in Europe. Although each additional cancer fatality represents a personal tragedy it must be said that the most likely number of cancers induced in the UK of, say, 400–500 over the next 40–50 years will be indistinguishable in the 6–8 million cases which will occur anyway in the population in that time. Even responsible arguments over the magnitude of this health effect do reduce the public's confidence in the scientific 'establishment'. However, all of this is in stark contrast to a statement made by the Secretary of State for the Environment, Kenneth Baker, on 6 May. 'The effects of the cloud have already been assessed and none presents a risk to health in the United Kingdom'. Nothing did more to increase public anxiety than misleading statements of this kind especially when it was followed only six weeks later by a ban on the sale and consumption of lamb because of persistence of ^{137}Cs in the environment.

The apparent ill-preparedness of the government in respect of not only the incident itself but also the behaviour of the fallout in the environment gives cause for worry. To start with, effective emergency plans should have been formulated to deal with a reactor accident occurring anywhere in Europe. We have now accumulated some 5000 reactor years experience world-wide and, as several risk studies have predicted core meltdowns at the rate of 1 in 10 000 reactor years (with confidence limits of 1 in 3000 to 1 in 100 000), this accident cannot be said to be entirely unexpected. Predictions as to behaviour of caesium in the environment give no more confidence. During and following the testing of weapons in the atmosphere during the late 1950s and early 1960s a number of ecological studies were done resulting in models of food-chain pathways for a variety of isotopes, notably ^{137}Cs. Drawing on this information and bearing in mind the increased deposition in high pastures in North Wales and Cumbria it should have been predicted that increased environmental retention would result in unacceptable levels of ^{137}Cs in lamb. Whether anticipation of this increased retention would have helped the farming community or even whether the ban was justified, is a matter for conjecture but the necessity for a policy of openness and full discussion of the issues resulting in such decisions is not.

To illustrate the societal-scientific problems associated with decisions made following a reactor accident such as the Chernobyl incident consider the following.

ENVIRONMENTAL CONTAMINATION

Assuming that environmental contamination is either predicted or totally unexpected, who provides an early warning? In the UK we seem to have a policy that the polluter does the monitoring or at least in the nuclear power field. The Ministry of Agriculture, Fisheries and Food (MAFF) do provide routine surveillance but this is seen by the public as the government monitoring its own utilities. Given that adequate monitoring on a national scale is complex, requiring considerable scientific expertise and capital investment in low-level radiochemistry and counting equipment, governmental involvement is not so surprising. However, there is also a strong argument for a network of remote sensors to detect variations in local gamma background and some involvement of local authorities in, at least, primary monitoring. One can sense and understand the frustration of local authorities, in the aftermath of Chernobyl, being unable to provide *local* information to the public on levels of contamination. A grand plan for environmental surveillance must include an element of *independent* data acquisition in which both hospital nuclear medicine departments and universities have a part to play. It has already become apparent that local authorities, particularly those with nuclear establishments within their boundaries, are concerned about monitoring and it would seem appropriate for the government to encourage and coordinate this on a national scale. However, the interpretation of the data so acquired is not a simple exercise and not the sort of task in which local authorities could have a role.

CONTAMINATION OF PEOPLE AND DOSES RECEIVED

Following the Chernobyl incident, as a result of inhalation of volatile isotopes and dusts and the intake of contaminated food and water people will have received internal radiation doses. Whether restrictions should have been placed on the sale of contaminated food and milk or whether thyroid blocking tablets should have been issued, is a value judgement which is, to some extent, a political rather than scientific issue. There is no doubt that because of inactivity on the part of the government some thyroid cancers will occur that could have been prevented—but at what cost? Throughout Europe following Chernobyl three factors have been considered by governments:

(i) To limit the health risks from radiation.
(ii) To calm people's concern and prevent panic.
(iii) To keep the domestic political and economic damage to a minimum.

Table 1 Intervention levels in Europe for ^{131}I in milk

	Milk (Bq/litre)
IAEA	
Children	1 000
Adults	10 000
WHO	2 000
EEC	
before 6 May	500
before 16 May	250
before 20 May	125
UK	2 000
Luxembourg	500
Sweden	2 000
France	3 700
Italy	5 550
The Netherlands	500
FRG	
Federal	500
Hessen	20

Figures assembled by WISE, Amsterdam.

Obviously at some stage a balance has to be struck between acceptability of risk and other considerations. These other considerations are usually rather vague assessments of benefit but in this case could include potential panic. For instance, even with hindsight, the widespread issue of iodide or iodate tablets may have caused more damage than it would have prevented. However, consistency over why actions are being taken or not is important and it was difficult for the government to explain why 'action' or intervention levels over the contamination of food varied so much throughout European countries (see Table 1).

Nevertheless, interpretation of para. 74 of ICRP 29[3] ['No single set of values (of intervention levels) can be provided because the choice of action will depend on a balance between the risks associated with the predicted doses or dose commitments and the risks associated with countermeasures'] might make it almost inevitable that action levels would depend on politics rather than science.

RISK AND PUBLIC PERCEPTION

The step in any assessment of health and detriment from individual or collective dose to risk is a major one and is the subject of considerable controversy. Public perception of risk is important here because the Chernobyl incident has demonstrated that there is an extreme (almost irrational) fear of radiation risks. Nevertheless there has to be acceptance of some risk associated with all man's endeavours and these risks must be estimated and compared as far as possible. No energy generating system is completely without risk but unfortunately the risk seems to be borne disproportionately by different groups of the population. For instance, for the generation of nuclear power the people of West Cumbria seem to bear the brunt of the risk. The risks of catastrophic accident are, of course, on quite a different scale from say coal burning and nuclear generation whereas they are much more comparable under normal conditions. In fact, under normal operating conditions it has been estimated[4] that in terms of all injuries the coal cycle is about seven times as hazardous as the nuclear fuel cycle. However, this estimate as with others[5] depends on radiogenic risk estimates derived from the ICRP recommendations[6] and there is now some evidence which suggests that the ICRP may be underestimating the risk by at least a factor of five.

UNCERTAINTIES IN RISK ESTIMATES

The strongest evidence of radiogenic cancer risk still comes from the survivors of the bombings at Hiroshima and Nagasaki although these could be considered to be a selected population and therefore atypical. However, in the absence of better data these have still to be used, and the numbers of cancers are still increasing. These data were used in the derivation of risk rates published in the BEIR III report[7] and the models used in this report seem to fit the more recent data. Because all people who were exposed have not died in any population studied the complete expression of cancer is uncertain and models have had to be used to extrapolate from known recorded incidences of cancer to a lifetime's experience. Two models have been employed; (i) the relative risk model which assumes the exposure to radiation multiplies the natural risk to an individual by some factor that stays constant throughout life although the risk factor will vary with age at exposure and (ii) the absolute risk model which assumes that the radiation exposure adds an annual risk that remains constant thereafter. ICRP[6] in making recommendations for setting standards used the absolute risk model whereas subsequent data have fitted a relative risk model. This has been endorsed by the US National Institutes of Health[8] and recognized now by the ICRP[9] although there are some fundamental objections. In the case of both lifelong exposure and single exposure the BEIR III committee estimated that the relative risk model would predict about 2.5–3 times as many deaths as the absolute model. This has also now been acknowledged by ICRP[9], although their recommendations of 1977[6] must therefore be an underestimate.

Many assessments of health detriment refer only to death from induced cancer, whereas the mere induction of cancer may in itself be traumatic. Total incidence was considered by the BEIR III committee, and ICRP acknowledges in Publication 45, p. 31[9] that there would be twice as many non-fatal cancers as fatal cancers induced by radiation.

Thus, these two factors alone suggest that ICRP risk factors may underestimate the incidence of radiogenic cancer by a factor of about 5. If the probable outcome of the revision of the dosimetry at Hiroshima and Nagasaki is to alter cancer risk rates derived from these exposures upwards by about 1.5, there is obviously an even stronger case for revision of dose limits (and, therefore, secondary derived limits). It is noteworthy that because of uncertainties of risk and in dose determination the public dose limit for continuous exposure has been dropped by a factor 5 to an average of 1 mSv/year. It is also noteworthy that, even without the data from the Japanese, risk estimates are thought to be underestimated by a factor of about 4.[10] In the face of this uncertainty it would seem prudent to err on the most conservative side and reappraise the dose limits. This at least would be one way of gaining some public confidence after Chernobyl.

CONCLUSIONS

There can be no doubt that the incident/accident at Chernobyl will result in an (undetectable) increase of cancer in the UK over the next 40–50 years—probably 400–500 cases occurring, less than half of which would be fatal. Preventive measures which could have been instigated would have had doubtful effects in terms of total health detriment in the population.

The overall response of the government to the incident was marked by confusion and an uncertain chain of responsibility. The need for independent monitoring and assessment, and effective communications with the public was highlighted by the number of unofficial bodies and 'instant experts' who were approached for advice and information on health effects.

There is an obvious need for some consensus about intervention or action levels among the countries of Europe. It has become clear that accidents of this nature will not respect national frontiers but international consistency of action will at least inspire some confidence.

Lastly, this accident has, I hope, emphasized the need for effective emergency plans not just for the unlikely event occurring here but the statistically greater chance of the next 'Chernobyl' occurring in, say, France.

REFERENCES

1. F. A. Fry, R. H. Clarke and M. C. O'Riordan, *Nature (London)*, **321**, 193–5 (1986).
2. F. von Hippel and T. B. Cochran, *Bull. of the Atomic Scientists*, 18–24, August/September (1986).

3. International Commission on Radiological Protection, Publication 29 (1979), Pergamon Press.
4. K. A. Hub and R. A. Schlenker, in *Population Dose Evaluation and Standards for Man and His Environment*, IAEA-SM-184/18 (1974), pp. 463–82.
5. A. V. Cohen *J. Soc. Radiol. Prot.*, **3**, 9–14 (1983).
6. International Commission on Radiological Protection, Publication 26 (1977), Pergamon Press.
7. Report of the 'Biological Effects of Ionizing Radiation' Committee of the US National Academy of Sciences and National Institutes of Health, (BEIR III Report) (1980).
8. Report of the National Institute of Health Ad Hoc Working Groupo to Develop Radioepidemiological Tables. *NIH Publication 85-2748*, US Department of Health and Human Sciences, Washington, (1985).
9. International Commission on Radiological Protection, Publication 45 (1985), Pergamon Press.
10. M. W. Charles, P. J. Lindop and A. J. Mill, in *Biological Effects of Low Level Radiation*, IAEA-SM-266/52. IAEA (1983), pp. 61–75.

Part 6
Conclusion

Radiation and Health
Edited by R. Russell Jones and R. Southwood

22
Concluding Remarks

SIR RICHARD SOUTHWOOD, FRS
(*Conference Chairman*)

At this conference we have heard a wide range of opinions on the issues before us and also the results of a number of scientific investigations whose conclusions, I perceived, were not as far apart as some of the opinions. Nevertheless, I hope that you will agree that any attempt by me to evaluate in detail the different scientific contributions—or reach firm overall conclusions—must be foredoomed to failure. Rather I will attempt to go over the areas discussed and identify some common ground, while highlighting areas of continuing uncertainty. It is of course on these that we must focus our research and evaluation in the immediate future.

Going back to the point I made at the start of the meeting, on one hand we have a potential cause and on the other an effect—its putative detriment on health. Whether we attempt to investigate the link between these through sequential systems models or through epidemiological analysis—clearly both cause and effect must be properly measured. Although only γ-radiation can accurately be measured in the body, in the environment all forms of radiation can now be measured with great precision. Therefore it should be possible to know exactly the levels of radiation to which persons are exposed. However, as both Drs Clarke and Lambert have said, we need a wider monitoring network. I agree with them and I would suggest that the Environmental Health Officers of local authorities should play a major role—as well as universities, polytechnics and nuclear establishments. A mixed network of this type would I believe provide an important element of public reassurance and, given the resources, I am sure that NRPB would be able to undertake the role of co-ordinator.

In assessing long-term effects it is past exposures that need to be determined. Professor Radford has drawn attention to revision of the estimates of radiation exposure suffered in the two atomic bomb explosions in Japan. His report is of great interest; as I understand it, the full data are being analysed by the Japanese authorities and UNSCEAR and I presume, and hope, that their conclusions will soon be public. Dr Lambert has shown some of the difficulties in extrapolating these curves.

We must await these formal reports before passing final judgement, but in the light of the foretaste Professor Radford has given us it seems that they will modify our understanding.

Professor Berry has outlined to us the history and approaches of ICRP who play such a major international role in advising on dose/effect relationships. Professor Berry showed how that body was a pioneer in the field of worker safety and how standards have improved with time. This is not unique to this field. Looked at historically there is really nothing absolute about standards in safety or environmental protection—any more than there is a unique level for living standards. To some extent the two are related—risk and opportunity tend to be traded off against each other. Mr Gee reminded us that society must ensure that such trade-offs are fair. However, I am digressing towards a discussion of 'acceptable risk'. Returning to the ICRP risk estimates, Dr K. Z. Morgan chides that body for some past decisions and for the speed—or rather as he perceives it the lack of speed—of its decision-making process. I hope, with several speakers, that the biological lessons of Chernobyl will be eventually incorporated into the ICRP's thinking though this must be far into the future. However, if we are not in the future to get into some of the difficulties currently experienced with the Japanese data, it will be necessary for our Russian colleagues to make very careful assessments of dose exposures over the last six months. Beyond the boundaries of the USSR we do have a much clearer perception as Dr Clarke's chapter shows. His survey and other data indicate that the effective doses though not to be diminished were such that the putative effects will not be detectable among the temporal and regional variation of the observed cancer frequency in the population. The deposition was extremely non-uniform, being greatly influenced by rainfall patterns; furthermore, as Dr Clarke stressed, each accident has different features depending on various physical and biological factors.

The measurement of variations in the distribution of radiation exposure is essential to further understanding—as most of our conclusions must be based on observation rather than experimentation. This radiation stems from a variety of sources and reaches us by several pathways as described by Miss Fry and Mr Taylor. Clearly it is important, when models are built using such inputs, to be able to test the predictions by monitoring actual levels. As mentioned by Dr Lambert and in the discussion, postmortem studies could be particularly valuable in testing some sections of such models though they have proved to have a number of practical and ethical difficulties. Notwithstanding the extent to which various components of current models have been tested, Mr Crouch has shown that they are not as robust as is desirable. Certainly few, if any, would claim that these models are exact—Dr Stather has prefixed some of his estimates by 'about'. The argument is whether the risk estimates are about right, with an error of about $\times 2$, or out by perhaps an order of magnitude as Professor Radford would suggest (because the detrimental effects of radiation have been underestimated) or out by more than the two orders of magnitude that Dr Stather has shown are needed on his model if the excess of leukaemias in Seascale are to be entirely accounted for by the known releases of radionuclides from the reprocessing plant. It is I think important to remember, as Dr Darby has stressed, that the expected number of cases is based on the effect of all radiation and that the artificial component, whenever it has been measured, is, as Dr. Stather showed, only a small proportion of that. A relatively small readjustment of dose–risk function will not account for the discrepancy at Seascale. Personally I would agree with those who suggest that if these excesses, these clusters, are to have a causative relationship with ionizing radiation then either our knowledge of the actual releases is out by a very large factor—or there is something about the way particular radionuclides behave that is outside our present understanding. If that understanding does change I think this is most likely to concern α particles—which are so difficult to track in the body. This is something that has been emphasised at this conference.

Clusters of disease are a cause of concern: what is the agent that leads to a greater probability of illness in a particular site? A cluster is 'a closely grouped series of events or cases of a disease—with well defined distribution patterns in relation to time or place or both'.[1] In some studies, such as those of Mr Urquhart, cases were defined geographically according to date and place of diagnosis; in her studies Dr Stewart worked on date and place of birth. Any differences in parental mobility, for example, could lead to divergence between findings with such differences in definition.

Everyone appreciates that a cluster could arise by chance—one can calculate the odds against that as Mr Urquhart pointed out, but no one would participate in lotteries or pools if they did not recognize that long odds may come up. Mr Urquhart has effectively said you may have one cluster by chance, but if there are several, is the dice loaded in some way? He pointed out that this should be investigated by making comparisons with control sites, that the situation in localities other than those around nuclear installations should be studied. That is one reason why I found Dr Stewart's chapter so interesting—she has looked across a wide geographical area—the whole country—and found 517 clusters of childhood cancers, obviously too numerous and too widespread to be associated with nuclear power stations. Is there some infective agent? Dr Darby has shown how, if we consider the effects of radiation from the fallout from nuclear tests as a standard, the cluster of leukaemia around Dounreay is not easily explained by radiation.

She suggested other agents that might be involved. When comparisons are made between urban and rural situations, we must remember that the construction of a large nuclear installation alters a rural environment in several ways apart from possible increases in radiation. Yet these clusters remain a challenge to us, and a cause of anxiety, until they are understood.

The chapters by Dr Stewart and Dr Ujeno also focused on the influence of natural background radiation. There are of course always pitfalls in simple correlative analysis and multiple regressions, but many studies do suggest that certain cancers are more likely to arise from radiation exposure than others—a point also made by Dr Beral.

So my review ends—where I might well have started—on the variation within the system. Clearly we know a great deal about radionuclides in the environment, about pathways to and in the body, about the doses delivered and the potential for cytogenetic damage. The broad outlines are largely agreed, but there can be no doubt that there are some inconsistencies—to interpret these we need to consider the components of variation. This is another general conclusion I would draw—the importance of paying particular attention to the variation—its form and causes. However strong the theory looks, in the face of a conflicting piece of evidence we must be prepared to look at each component again. Is there any possibility that actinides become distributed in the body so that the received dose is larger than believed? It seems not—*but we must keep looking*. Then there is the question of different forms of radiation; ICRP has given and continues to give much attention to this.

The next step is the response at the cellular level. As Professor Evans has reminded us, mutagenic damage may arise from many causes. There is certainly evidence[2] that leukaemias may be caused by a viral organism. It is also known in other situations that there can be synergistic effects between some physical or biological stress and the expression of a viral illness. There is some evidence for radiation releasing mutagenic agents[3]; sunburn will release herpes virus. Thus this is one of the possible links we must always be on the lookout for if we are to discharge the duty on so many here of advising (officially or unofficially) on radiological protection.

The last variable to which I want to refer is the variation between individuals. Dr Lewis and Professor Evans have drawn attention to this and following up my point about synergism above I will remind you that Dr Lewis indicated that certain rare diseases may greatly increase the sensitivity of patients to radiation burns. Variation may of course occur not only between different individuals, but also within the same individual at different ages—as Professor Radford and Dr Stewart have pointed out—and Dr Stewart's own earlier work indicated, younger persons and especially prenatal persons may be more sensitive to certain exposures and more liable to certain diseases.

At the end of the day though, regulatory agencies have to set certain standards for public protection and Dr Clarke illustrated some of the complexities in this procedure in relation to emergencies. The more we understand about the variables we have had drawn to our attention at this conference, the more confident we can be in making recommendations to reduce risk. This can of course never be zero and the natural component will ensure that.

I will emphasize again my own belief that we know and understand a lot and we are certainly not likely to be so disastrously wrong that non-medical radiation is killing as many persons as, for example, die from traffic accidents or even sun-induced melanomas. Dr Beral's findings sustain this view. But we must not be complacent—the uncertainties and inconsistencies that exist pose a challenge and if and when research provides new evidence we must be prepared to change our thinking to ensure that, we—all of us interested in this subject—have provided the correct estimate of risk. Finally we return to the question that I placed outside our main agenda—what is an acceptable risk? This is a very complex process, more a subject of political science than biological science—for it involves the balancing of public concern, economic benefits and costs and scientific evidence—Mr Gee's contribution, highlighted this. As the studies of the Royal Society[4] and others have shown, public concern about a risk is never simply expressed as the mathematical probability of something occurring; risks willingly undertaken (like driving a car) are more readily accepted than those arising from the decisions of others.[5] Then, as has been pointed out elsewhere (e.g. Southwood,[6]) and Mr Taylor emphasized here, in issues such as this scientific evidence is rarely 100 per cent conclusive: one has a level of confidence—not a formal statistical level—but something analogous. Of course it is much easier for those who have to make the decision if the scientists are agreed and completely confident about an effect—as we are about the effect of DDT on birds of prey.

But when the conclusion is at a level of uncertainty between 'it is possible'—to 'it is very likely'—then it is necessary to consider the cost and implications of a decision either way. If the view that the risk is real is accepted and in the end this turns out to have been a mistake — how great is the cost of this error — and there are both costs to those whose investment may have been rendered nugatory and costs to society in general that has, for instance, missed an opportunity for cheaper transport. There is also the converse—if the risk is disregarded and then proved real—what will be the costs and how quickly can the risk be attenuated. Let me illustrate the former by two examples. There is some evidence that CFCs affect the ozone layer, the uncertainty is considerable and the scientific consensus has varied over the last decade[7]—but why not ban their non-essential use in spray-on cosmetics? Roll-on applications are quite adequate. My view is therefore that as the costs to society are so few, action would be justified on the present relatively slender evidence. Another example is lead in petrol; I will not repeat the story in detail. There are several reasons for considering the supposed health risks from lead as very likely to be real.[6,8,9] Although the costs, in the widest sense, of doing without lead for many of its uses are greater than those for the loss of CFC in non-essential uses—the balance is still clearly in favour of accepting that lead is harmful. By phasing out its use, which is present UK policy, the investment costs to industry can be minimized.

We will all have our individual judgement as to where the balance lies with radiation risks and benefits. I referred to the balance of cost and benefit in its medical use in my opening remarks. Another major use is in energy production and Dr Lambert pointed out that the risks associated with the use of fossil fuels are probably less than those using nuclear power. For my part I cannot see the world significantly reducing its energy demands in the near future and, recognizing the environmental disbenefits of all forms of electricity generation, I believe we will need a diversity of sources. However, whatever our current view on this balance, this conference will have made an important contribution towards enabling us to be better informed. Many different views have been expressed, but all seem to have taken the attitude expressed so pithily by Professor Berry in the discussion 'I hear you'. I trust we can go on hearing each other.

REFERENCES

1. J. M. Last, (ed.), *A Dictionary of Epidemiology*. (IEA & WHO Handbook) Oxford University Press, 1983, 114 pp.
2. M. S. Linet, *The Leukaemias: Epidemiologic Aspects*, Oxford University Press, 1985, 293 pp.
3. G. E.Adams, Radiation carcinogenesis, in L. M. Franks and N. M. Teich (eds) *Introduction to the Cellular and Molecular Biology of Cancer*, Oxford University Press, 1986, pp. 154–75.
4. Royal Society, *Risk Assessment*. Report of a Royal Society Study Group. Royal Society, London, 1983.
5. S. Jasanoff, Risk Management and Political Culture. *Social Research Perspectives Occasional Reports on Current Topics*, **12**, 1986 93 pp.
6. T. R. E. Southwood, The role of proof and concern in the work of the Royal Commission on Environmental Pollution. *Marine Pollution Bulletin*, **16**, 346–9 (1985).
7. RCEP (Royal Commission on Environmental Pollution), IXth Report. *Lead in the Environment*, HMSO Cmnd 8852. London, 1984, 184 pp.
8. RCEP (Royal Commission on Environmental Pollution), Xth Report. *Tackling Pollution-Experience and Prospects*, HMSO Cmnd 9149. London, 1983, 233 pp.
9. T. R. E. Southwood. Lead in the environment, in G. R. Conway (ed) *The Assessment of Environmental Problems*. Imperial College Centre for Environmental Technology, London, 1986, pp. 11–22.

Index